GOOD GOVERNANCE AND SUSTAINABLE DEVELOPMENT

Addressing the big questions to advance the debate on good governance for sustainable development, this book asks: what is governance for sustainable development? What makes governance 'good'? And what are the factors that inhibit good governance?

With contributions from experts in Development Studies, Economics, Geography, Management, Environmental Science, Politics, and Public Policy, this interdisciplinary volume critically engages with the Sustainable Development Goals (SDGs) whilst reflecting on the inherently political nature of good governance. Chapters draw on cases from a range of geographical contexts (local, national, supranational, and global) and diverse policy areas, including global environmental and labour governance, gender equality, decolonial governance in the Coral Triangle, supranational governance in the EU, and resource governance in the car industry. By doing so, this book provides an overview of the current challenges and possibilities for achieving good governance and sustainable development, and calls for a return to the critical spirit of the original concept of good governance, to reclaim it as the radical idea it was intended to be.

This book will act as an insightful and valuable guide for academics, scholars, and students of Development Studies, Economics, Politics, Environmental Science, Geography, and Management, as well as policymakers and practitioners working in sustainable development.

Maria Gavris is Associate Professor in the Department of Global Sustainable Development, School for Cross-faculty Studies at the University of Warwick, UK. She is a political economist with a particular interest in governance for sustainable development, capitalism's multiple interconnected crises, and alternatives to the mainstream development paradigm. At Warwick, she designed and convenes an interdisciplinary module on Good Governance and Sustainable Development.

Jessica Savage is Associate Professor in the Department of Global Sustainable Development, School for Cross-faculty Studies at the University of Warwick, UK. She is an interdisciplinary environmental scientist with a background in tropical coral reef ecology and natural resource management. Her work is focused on the efficacy of marine protected areas and their impacts on the environment, economy, local indigenous communities, and tourism. Her research interests surround human-ecosystem interactions, and in particular the monitoring of marine resources.

Routledge Studies in Sustainable Development

This series uniquely brings together original and cutting-edge research on sustainable development. The books in this series tackle difficult and important issues in sustainable development including: values and ethics; sustainability in higher education; climate compatible development; resilience; capitalism and de-growth; sustainable urban development; gender and participation; and well-being.

Drawing on a wide range of disciplines, the series promotes interdisciplinary research for an international readership. The series was recommended in the *Guardian*'s suggested reads on development and the environment.

How to Achieve Sustainable Development
Local Actions for Global Change
Samuel T. Ledermann

Innovation Strategies and the Sustainable Development Goals
A Blueprint for Sustainable Transformation
Amit Singh Khokhar, Punita Duhan and Anshu Lochab

Artificial Intelligence for Sustainable Development
How, Why, and What Drives the Future
Medani P. Bhandari

Good Governance and Sustainable Development
Edited by Maria Gavris and Jessica Savage

For more information about this series, please visit: www.routledge.com/Routledge-Studies-in-Sustainable-Development/book-series/RSSD

GOOD GOVERNANCE AND SUSTAINABLE DEVELOPMENT

Edited by
Maria Gavris and Jessica Savage

Routledge
Taylor & Francis Group
LONDON AND NEW YORK

Designed cover image: Shutterstock

First published 2026
by Routledge
4 Park Square, Milton Park, Abingdon, Oxon OX14 4RN

and by Routledge
605 Third Avenue, New York, NY 10158

Routledge is an imprint of the Taylor & Francis Group, an informa business

British Library Cataloguing-in-Publication Data
A catalogue record for this book is available from the British Library

ISBN: 9781032741093 (hbk)
ISBN: 9781032741116 (pbk)
ISBN: 9781003467694 (ebk)

DOI: 10.4324/9781003467694

Typeset in Optima
by codeMantra

CONTENTS

ILLUSTRATIONS

Figures

Tables

CONTRIBUTORS

Vanessa Burns is a Teaching Associate at the Centre for the Study of Existential Risk at the University of Cambridge where she researches extreme climate change and teaches on the MPhil in Global Risk and Resilience. Vanessa's work is broadly concerned with how environmental knowledge is produced in law and governance frameworks, and how this obstructs good environmental governance. She is especially interested in human adaptation to environmental change, adaptation injustice, and how societies are preparing for climate change beyond 2030. Vanessa has led a number of international environmental research projects on climate change, environmental governance, and food and water security. She holds a DPhil in environmental geography from Oxford, is an Honorary Research Fellow at the University of Sheffield, a Fellow of the Royal Geographical Society, and an Affiliate of the Centre for Integrated Research in Risk and Resilience at Kings College London.

Dan Coffey is an economist and writer. He is the author of over 70 publications and productions, including *The Myth of Japanese Efficiency: The World Car Industry in a Globalizing Age* (2006) and (as co-author) *Globalization and Varieties of Capitalism: New Labour, Economic Policy and the Abject State* (2009). His most recent co-edited collection is *Global Economic Crisis and Local Economic Development* (2016), published by Routledge. He is on the Steering Committee of GERPISA ('the international research network of the automobile'). His current interest in the car industry focuses on the challenges of an environmentally sustainable outcome for the car.

Anastasia Ershova is Assistant Professor and Ad Astra Fellow at University College Dublin, Republic of Ireland. Her research interests include EU policy- and decision-making, policy responsiveness and public opinion, and delegation choices in the supranational arena. Prior to joining UCD, Anastasia worked as a Lecturer in Public Policy at Queen's University Belfast.

She was also employed as a Post-Doctoral Researcher in a NORFACE project 'EUINACTION' at Leiden University, the Netherlands, and as an LSE Fellow at the London School of Economics and Political Science, London, UK.

Marta Guerriero is Reader and Head of Department in Global Sustainable Development at the University of Warwick. She is an applied economist working at the intersection between development, environment, and labour economics. Her research is primarily focused on issues of socio-economic justice and policy evaluation. She approaches her research and teaching in an interdisciplinary manner, also incorporating aspects of political economy, labour and development studies, and environmental and social policy. During her career, she has taught several undergraduate and postgraduate modules in global sustainable development, and she has worked with a number of private and public organisations in the field.

Jason Heyes is Professor of Employment Relations at the University of Sheffield and the Director of the Sheffield University Management School's Centre for Decent Work, UK. His research interests include labour market transitions, underemployment, and labour governance. His research has been published in journals including the *British Journal of Industrial Relations*, *Human Relations*, and *Work, Employment & Society*. He has worked regularly with the ILO on research and technical projects related to labour administration and has co-edited two books on labour administration, both of which were published by the ILO and Edward Elgar.

Molly Wimonmat Srichamroen is Lecturer in Public Management and Public Policy in the Department of Public Administration, Faculty of Political Science, Chulalongkorn University, Thailand. She holds a PhD in Public Policy from Victoria University of Wellington, New Zealand. Her research interests include sustainability policy and management in the public and private sectors, strategic planning, policy solution designs, service delivery improvement, and social policy.

Carole Thornley is Emeritus Professor of Employment and Public Policy at Keele Management School at Keele University. She has authored around 80 books, articles, and commissioned studies, and is co-author of *Globalization and Varieties of Capitalism: New Labour, Economic Policy and the Abject State* (2009) and co-editor of *Globalization and Precarious Forms of Production and Employment: Challenges for Workers and Unions* (2010) and *Global Economic Crisis and Local Economic Development* (2016). Her research interests include globalisation, industrial organisation, industrial strategy, employment, public policy, and productivity. She is on the Steering Committee of GERPISA ('the international research network of the automobile'). She advises trade unions, and has had a long association with the British public sector union, UNISON.

ACKNOWLEDGEMENTS

This book has truly been a labour of love. The original idea took shape in corridor conversations (our offices share a wall) and grew out of the modules we teach. Generations of Global Sustainable Development students at Warwick have played a crucial role in shaping and challenging our thinking: if you are reading this, know that this work exists in large part for you, and because of you.

We are deeply grateful to our editorial team at Routledge, who supported this project from the beginning and whose expertise was essential in bringing this book to publication. Our contributors have made this book what it is today, and we are very thankful for their enthusiasm for the project and the insights they brought to it.

Finally, our heartfelt thanks go to our families for their unwavering love, care, and support. Our husbands, Joe Davidson and Tom Savage-Wilkes, deserve special recognition for their patience and encouragement throughout this journey, as does Henry Savage-Wilkes, who was born and grew alongside this book.

Maria Gavris and Jessica Savage
Coventry, November 2025

1

INTRODUCTION

Maria Gavris and Jessica Savage

Introduction: aims and significance of this book

Good governance is widely regarded as a fourth necessary pillar of sustainable development, in the absence of which no progress can be achieved on any of the other three pillars – environmental, social, and economic. As long ago as 1998, then UN Secretary-General Kofi Annan stated that 'good governance is perhaps the single most important factor in eradicating poverty and promoting development' (UN, 1998, 13). By the turn of the 21st century, 'a significant portion of the development agenda was related to good governance' (Grindle, 2012, 259). More recently, intersecting global crises in the environmental, economic, and social realm, collectively referred to as the 'polycrisis' (Tooze, 2021), have highlighted the urgent need for governance reforms (Haifa, Zineb, and Ziad, 2023). Despite its centrality, however, the concept of good governance in the context of sustainable development remains ambiguous and under-researched, beyond the general notion in the literature that institutions matter for development in some way. But what do we mean by good governance? What constitutes sustainable development? What is the relationship between the two?

In practice, instead of encouraging strong and meaningful debate on the topic, the dominant interpretation of 'good governance' in contemporary discourse is heavily shaped by the World Bank's sustainable development agenda. This framework equates sustainable development with economic growth, neglecting its social and environmental dimensions. It reduces 'good governance' to a set of numerical indicators, often lacking robust theoretical foundations, which are tied to conditions for financial aid to developing countries. Overall, the World Bank's version of 'good governance' has been

DOI: 10.4324/9781003467694-1

criticised as a neocolonial project rooted in market fundamentalism, with significant distributional implications (see e.g. van Arkadie, 2012; Adejumo-Ayibiowu, 2015; Haifa, Zineb, and Ziad, 2023). So pervasive is the World Bank version of the term, that over time more radical and alternative interpretations of good governance have been sidelined. The central argument of this book is that while good governance does indeed matter for sustainable development, it must be understood in a different way than is commonly accepted today. We advocate for reclaiming the concept's critical edge and transformative potential.

Our role as educators was our primary inspiration for this book. We both teach on undergraduate and postgraduate interdisciplinary programmes in Global Sustainable Development at the University of Warwick. Governance issues are at the core of our modules GD216 Good Governance and Sustainable Development and GD317 Managing Natural Resources. However, we found it challenging to identify an accessible text on these themes that we could recommend to our students. The limited reading resources in this space either adhere to the narrow conception of good governance promoted by the World Bank or, if they diverge from this view, focus solely on specific regions or categories of states (e.g. South-East Asia, Western democracies), rather than offering a global perspective. A further issue is that much of the recent literature on the topic treats sustainable development as synonymous with the United Nations Sustainable Development Goals (SDGs). This is an issue, given both their upcoming expiry (2030) and the range of criticism directed at the SDG framework elsewhere in the sustainability literature. For instance, critics have highlighted the framework's failure to account for power relations at various levels of the global system and the obstacles these pose to achieving sustainability (Esquivel, 2016), the hierarchical-yet-non-hierarchical nature of the SDGs, and the framework's failure to acknowledge the interactions between sustainability goals (Kotzé, 2018), as well as the tendency to subordinate the Earth system to the dictates of the economy (Adelman, 2018). And while there are valuable books on specific sub-areas of governance (e.g. environmental governance), there is a lack of holistic, interdisciplinary texts on good governance for sustainable development that simultaneously address environmental, social, and economic governance, as well as the normative implications of the epithet 'good.'

We therefore set out to write this book to address some of these gaps and provide an accessible point of entry into the topic. This volume aims to advance the debates on good governance for sustainable development as an underexplored area, by bringing together experts from different disciplines (Development Studies, Economics, Environmental Science, Geography, Management, Politics, Public Policy) to assess the quality of governance, definitions of 'good,' and changes needed and challenges faced in order to achieve sustainable development in a wide range of geographical areas and policy contexts, and at different (and often intersecting) governance levels. The goal of this book is thus to provide a range of alternative ideas and applications on what constitutes good governance for sustainable development, challenging the mainstream good governance agenda. Contributors reflect critically on the meaning of 'good governance' in the context of sustainable development. They consider questions such as: What makes governance 'good'? What are the factors that inhibit good governance? What institutional changes are required (at the local, national, regional, global systemic level) to achieve sustainable development? Based on these reflections, we as co-editors provide a novel multidisciplinary response to the question of what makes governance 'good' in a sustainable development context.

We are hoping that this book will serve as a useful starting point for thinking about what governance for sustainable development might and should entail, both for students and academics in the broad field of Development Studies and cognate disciplines. Given the strategic importance – globally, nationally, and locally – of the good governance debate, we expect this book's discussions around governance challenges and possibilities will also be of interest to policymakers and development practitioners.

The rest of this chapter provides a detailed introduction to governance for sustainable development, setting the stage for this book and establishing the conceptual foundations for a theory of good governance in this context. It includes a critical discussion of the core concepts used throughout this book: governance, good governance, and sustainable development, on which the subsequent chapters will build. This chapter concludes with an overview of this book and a brief summary of each individual chapter.

Good governance and sustainable development

What is governance?

Governance is a popular but elusive term, with numerous definitions that lack universal consensus. This ambiguity stems, in part, from the concept's interdisciplinary nature. Governance is best understood at the intersection between different disciplines and subdisciplines, such as politics, public administration, law, sociology, economics, and international relations. Due to its broad scope, it is often preceded by a qualifying adjective – such as 'global,' 'urban,' 'environmental,' or 'good' – to lend some precision. However, the epithet 'good' in particular adds complexity, as it introduces a normative, philosophical dimension.

Some of the most concise definitions of governance describe it as 'the institutional underpinnings of public authority and decision making' (Grindle, 2012, 261). It can also be seen as a theory of governing, where governing is defined as the 'totality of interactions, in which public as well as private actors participate, aimed at solving societal problems or creating societal opportunities' (Kooiman, 2003, 4). These definitions emphasise the central role of institutions in governance, as well as the connections between governance and societal objectives, but remain relatively vague beyond that.

Perhaps a slightly more focused way to understand governance is to consider how it differs from the notion of 'government' (Plumptre and Graham, 1999). Unlike government, which is confined to the national level, governance 'operates at every level of human enterprise, be it the household, village, municipality, nation, region or globe' (UNDP, 2011, 287). It is no coincidence that the term 'governance' has increased in popularity with the advent of globalisation and the rise of marketisation under neoliberalism, two processes which have redefined the role of the state. As Streeten (1992, 1) put it:

> The state has become too big for the small things, and too small for the big things. The small things call for delegation downwards, to the local level, to participatory organisations, to decentralised decision-making, and to the market. The big things call for delegation upwards, for coordination between national policies or for transnational institutions.

This is not to suggest that we have witnessed a complete 'retreat of the state' (Strange, 1996) in recent decades. Rather, these changes mean that the interactions and institutions involved in policymaking have multiplied and become more complex, still including the state but now also a range of non-state actors, from local organisations and NGOs to businesses and global institutions (Evans, 2012). In other words, governance is not only interdisciplinary, but also multilevel, encompassing the state as well as actors at levels above and below it.

Some definitions of governance draw on the term's Greek origins, with the Greek *kubernaein* meaning 'to steer' (as in, steering a ship or vessel). For instance, Plumptre and Graham (1999, 3) define governance as 'the art of steering societies and organizations.' This raises questions about the purpose of steering: what is it exactly that we are steering society towards? After reviewing a range of different definitions of governance, Gisselquist (2012, 4) identifies three common elements that frequently appear across these definitions, which collectively describe governance as '(1) the process (or manner) through which (2) power (or authority) is exercised (3) to manage the collective affairs of a community (or a country, society, or nation).' Like the definitions based on the notion of steering, this poses further questions about who holds power, how it is exercised, and to what ends, or – to put it slightly differently – what governance is actually for.

In a context where the literature, either implicitly or explicitly, suggests that governance is done with a specific purpose that must be identified, it is not difficult to see why governance has come to be associated with sustainable development. The latter appears to be a worthy goal to steer society towards. However, as will be discussed below, sustainable development itself is a complex and contested concept. Given this complexity, the relationship between governance and sustainable development is poorly understood (Gisselquist, 2012). It is worth noting here that some scholars even argue it is not governance that leads to sustainable development, but *vice versa*: sustainable development can drive improvements in governance (see e.g. Grindle, 2012).

What is governance for: steering society towards sustainable development?

The term 'sustainability' was first introduced within the field of forestry, where it referred to 'never harvesting more than what a forest yields in new growth' (Kuhlman and Farrington, 2010, 3437). Documented usages of the term with this meaning date back to the 18th century. Over time, it has acquired a much wider meaning, to denote maintaining forever (or, more realistically, for as long as possible) our ecological systems, both human and non-human life on Earth, welfare, and wellbeing (Costanza and Patten, 1995). For some, sustainability implies something static or constant, which they argue is incompatible with the dynamic concept of development, leading them to view 'sustainable development' as an oxymoron (White, 2013). However, another interpretation is that sustainability is a long-term goal or ideal, with sustainable development serving as the pathway to achieve it.

The most widely cited definition of sustainable development comes from the Brundtland Report (1987): 'development that meets the needs of the present without compromising the ability of future generations to meet their own needs.' On the one hand, the definition appears to connect human development and actions to environmental carrying capacities across generations. On the other hand, the rest of the report presents an anthropocentric view of 'needs,' prioritising economic growth over planetary health and interspecies justice (Kopnina, 2016; Kotzé, 2018). This is contradictory to earlier definitions of sustainability which focused on the preservation of life on Earth in the long term. In 2002, the World Summit on Sustainable Development became the first global forum to explicitly define sustainability and sustainable development in terms of three interconnected pillars: the environment, society, and the economy. The notion of governance as a fourth pillar became popular in the 2010s, in the context of discussions surrounding a Post-2015 Development Agenda, as reflected, for example, by the *International Workshop on Governance 'of' and 'for' Sustainable Development Goals* which took place in New York in 2014 (Kanie, Zondervan, and Stevens, 2014).

One interpretation of the three pillar approach is that it traces its origins to the counterhegemonic ecological movement of the 1960s and 1970s, which criticised the economic status quo from an ecological and social standpoint

(Purvis, Mao, and Robinson, 2019). This critique highlighted the need to rethink the role of the economy in society in relation to ecological limits, and proposed a model of economic development that was decoupled from growth and aligned with equity and environmental health. Such considerations show how sustainable development was once a radical idea aimed at the genuine betterment of society. However, with the rise of neoliberalism, the critical potential of the three-pillar model was distorted into the notion of weak sustainability: the idea that the pillars are so closely interconnected that they are, to some extent, interchangeable. This view suggests that environmental degradation caused by economic growth is not a concern, as long as the environmental value lost can be compensated by generating more economic value. 'Weak sustainability' influenced the Brundtland Report and subsequent UN publications, and as a result, sustainable development is now often reduced, at least implicitly, to economic growth, and used to legitimise neoliberal economic policy.

This problem is also evident in the SDGs, the present-day incarnation of sustainable development. The SDGs consist of 17 globally agreed goals, adopted at the UN's 70th anniversary meeting in New York in 2015, with a target for completion by 2030 (also known as Agenda 2030). These goals, broken down into 169 targets, carry the ambitious tagline 'Transforming our World' and are an update on the earlier Millennium Development Goals (MDGs), which expired in 2015. The MDGs were criticised for their paternalistic nature, as goals developed by the Global North for the Global South to implement in order to address poverty. The SDGs, by contrast, are designed for all countries, recognising that no nation in the world has achieved sustainability yet, and they have a wider focus, beyond just poverty alleviation, with goals aiming for gender equality, biodiversity protection, or quality education, to name a few. While the SDG framework is thus much broader than the MDGs, it still lacks a holistic, integrated approach. It fails to adequately address how the goals are interconnected and dependent on each other, particularly in terms of how none of the goals can be achieved in the long term without securing the necessary environmental foundations. Goals related to consumption, production, and economic growth – entirely dependent on the Earth's systems – should be ranked lower than those focused on preserving the ecosphere and combating climate change (Kotzé, 2018). Instead, the SDG framework 'subordinates the Earth system, most of the global population and all other species to the "needs" of the economy' (Adelman, 2018, 39). It conflates economic growth with societal progress, viewing growth as the primary means of reducing poverty and generating the resources needed to achieve the other SDGs (Adelman, 2018; Esquivel, 2016). This disregards ecological limits, planetary boundaries, and scientific evidence demonstrating the impossibility of decoupling economic growth from environmental degradation. Despite its lofty tagline, Agenda 2030 essentially supports business as usual, perpetuating existing power dynamics – the 'big elephant in the room of Agenda 2030' (Esquivel, 2016, 12). The SDGs fail to address the current and historical injustices imposed by the Global North, the Bretton Woods institutions, multinational corporations, and international financial institutions against the rest of the world – obstacles that stand in the way of achieving sustainability.

Collectively, these critiques suggest that the SDG framework is not only ineffective in promoting genuine sustainability globally, but may also exacerbate unsustainability by supporting the status quo and postponing important conversations about the critical and urgent governance actions needed to address the crises of the Anthropocene (Kotzé, 2018). In light of this, and with the SDGs set to expire in 2030 and no clear successor in sight, we believe this is a highly opportune and timely moment to contemplate alternatives to the contemporary development paradigm. Even if we agree that governance involves steering society towards sustainable development, the meaning of sustainable development itself is in flux and malleable. We encouraged contributors to engage critically with the contemporary mainstream development paradigm when considering governance for sustainable development in a broader sense.

Having explored the concepts of 'governance' and 'sustainable development,' and how the latter may be a way to provide purpose to the former, we can now engage with the normative dimension introduced to the governance discourse by the adjective 'good.' This adds complexity to an already complicated theoretical foundation, meaning there is no simple answer to the question of what makes governance 'good' in the context of sustainable development. However, we can identify a dominant line of thought on this issue, as well as more critical approaches that challenge this prevailing perspective.

What makes governance good? Mainstream and alternative approaches

Like governance itself, good governance remains a poorly specified concept (Gisselquist, 2012). In practice, it is often easier to pinpoint examples of 'bad' governance than to establish a clear definition of what constitutes good governance. Given the link to sustainable development, one might argue that governance is good if it leads to tangible development, that is, if its outcomes match its aims. However, as noted above, if the aim of sustainable development itself is ambiguous and contested, then we may end up with radically different interpretations of good governance. This reality that good governance is value-laden and subjective is insufficiently acknowledged, with one version of good governance unquestioningly presented as the norm, dominating all others.

The dominant version of 'good governance' in contemporary usage has been heavily influenced by a series of World Bank reports published in the late 1980s and early 1990s, which analysed Africa's development trajectory in the decades immediately following independence. The 1989 report expressed concern about the continent's future, declaring 'a crisis of governance' in Africa (World Bank, 1989, 60), caused by 'systemic failures' of the state (World Bank, 1989, 61). It portrayed African countries as plagued by corruption, oppression, and nepotism, which erode policy legitimacy and credibility, but suggested that these problems could be overcome by empowering the private sector, which would help restore democracy (World Bank, 1989). The report also positioned structural adjustment as essential for the development of African states. The 1992 report prescribed efficiency, accountability, and transparency as solutions to Africa's governance challenges (World Bank, 1992). These have since become key good governance indicators, in what Evans (2012, 6) describes as 'a checklist approach.' The World Bank formalised its definition of good governance through the creation of the Worldwide Governance Indicators which measure six dimensions annually for over 200 countries: (1) Voice and Accountability, (2) Political Stability and Absence of Violence/Terrorism, (3) Government Effectiveness, (4) Regulatory Quality, (5) Rule of Law, and (6) Control of Corruption. According to the World Bank (2024), improvement of these indicators 'helps countries increase economic growth, build human capital, and strengthen social cohesion.' Other global institutions, like the IMF, OECD, and later, the UN, followed suit in declaring good governance important for their activities, and modelling their definition of good governance on that of the World Bank. For the UN, good governance consists of eight elements: 'participatory, consensus oriented, accountable, transparent, responsive, effective and efficient, equitable and inclusive and follows the rule of law' (UN, 2009). This closely mirrors the mainstream World Bank approach, albeit with somewhat more concern for equity issues. Such indicators are designed to provide a framework that is easily understood and used by different development stakeholders, including for 'general cross-country comparisons of governance and for evaluating broad trends over time' (World Bank, 2024). In other words, their assumed advantage lies in offering a coherent and accessible model.

Some commentators (see e.g. Andrews, 2008; Massey, 2022) suggest that the main issue with this mainstream approach to good governance is its reliance on 'indicators without theory' (Andrews, 2008, 379). This approach is criticised as a mere tick-box exercise with no deeper theoretical justification, which fails to shed light on what actually makes governance good in practice and why: 'pressure to develop best practice models, measure performance, and create data has drowned out clear and consistent thinking about what was being done' (Andrews, 2008, 381). Additionally, it overlooks the possibility that governance may be good in different ways depending on context. It results in an unhelpful, static, one-size-fits-all model of good governance which amounts to 'telling developing countries that the way to develop is to become developed' (Andrews, 2008, 380). Factors such as demographics, revenue, or natural resource endowments constitute significant differences between countries at different levels of development, such that good governance may in fact look quite 'different in different developmental contexts' through time (Andrews, 2008, 381).

While this is an important critique, in this book, we take this further to argue that the problem is not so much a lack of theory as such, but rather the implicit theory and ideology surrounding both governance and sustainable development embedded in the dominant good governance agenda. As evidenced by the reports cited above, the World Bank's version of good governance is grounded in a particular interpretation of sustainable development as economic growth, and constructs economic liberalism as the precondition for democracy. This perspective serves to legitimise neoliberal capitalism – and structural adjustment policies (SAPs) specifically – and positions it as the natural state of affairs. The development vision promoted by this approach serves specific interests, yet it is presented under a veneer of impartiality, suggesting that the World Bank's idea of development is universally beneficial and shared by all.

As noted by Haifa, Zineb, and Ziad (2023), good governance as a concept – and later a discourse with its associated set of policies – was popularised by the Bretton Woods institutions at a time of US unipolarity. In this context, it became a hegemonic instrument wielded against underdeveloped nations. From an Afrocentric perspective, Adejumo-Ayibiowu (2015) argues that to uncover the theoretical foundations underpinning the good governance indicators, we must ask the question 'whose governance is this?' She concludes that the good governance agenda in its present form primarily serves the interests of the Global North and global neoliberal institutions. The conditions attached to loans under SAPs in the name of 'good governance' directly undermine democracy in the affected countries, instead of fostering it (Adejumo-Ayibiowu, 2015). In the words of van Arkadie (2012, 60), 'even if one personally shares these ideals, one can question whether wealthy States have the right to impose these values on poorer countries through development assistance, or whether this is not a form of neocolonialism.' Neocolonialism is also evident in the fact that good governance indicators and assessments are tied to the racist concept of 'state failure' (Gruffydd Jones, 2014; Icaza, 2022) and the assumption that Africa as a continent has failed and must be guided onto the right path by the West. But, as Adejumo-Ayibiowu (2015, 19) reminds us:

> African societies were not impoverished or badly governed before the advent of colonialism by the West. Most evidence of poor governance such as nepotism, patronage network, corruption and dictatorship was not inherent African nature; rather, those ills became pronounced in Africa due to colonisation and other Western influence.

At this juncture, given these problems, one might reasonably question whether it would be better to simply abandon the concept of good governance altogether. However, doing so would also mean discarding the concept's radical, anticolonial, and anti-imperialist roots and potential. The Malawian economist Thandika Mkandawire (2007) situates the concept within a broader historical context, documenting how the idea that Africa could benefit from good governance originally emerged among African scholars studying state-society relations in Africa during the second half of the 20th century. In this research, good governance meant something very different compared to the present-day good governance agenda. The African idea of good governance focused on issues of equity and inclusion in postcolonial states and sought ways to strengthen newly liberated African nations while reducing external influences. According to Mkandawire (2007), the World Bank, in preparing its 1989 report, consulted African scholars (including the Nigerian political philosopher Claude Ake, the Senegalese political economist Makhtar Diouf, and the Kenyan political scientist Ali Mazrui) on the main challenges facing their countries, and co-opted their concerns with good governance. However, the World Bank divorced the concept from its original intent, instead using it as a euphemism for policy conditionality. As Mkandawire (2007, 681) states, good governance as we know it today 'sidestepped the central concerns of the Africans and rendered the notion purely administrative.' Nevertheless, even once the World Bank approach became the dominant discourse, alternative and radical formulations of good governance continued to exist, though marginalised. For example, Arab and Palestinian definitions of good governance emphasise social justice, participation, and solidarity (Haifa, Zineb, and Ziad, 2023). In their study of indigenous groups in Canada, Plumptre and Graham (1999) found that their definitions of good governance differ profoundly from the mainstream, and are derived from indigenous traditions, histories, and cultural norms.

Thus, while the pro-growth and neocolonial undercurrents of the contemporary good governance agenda should not be underestimated, this does not mean that the broader idea of good governance is inherently flawed and should be rejected entirely. Rather, it is the dominant version of good governance in circulation that is problematic, but this is only one of many possible approaches. Grindle (2012, 259) maintains that 'good governance is a seductive idea,' because no one 'can reasonably defend bad governance.' Van Arkadie (2012, 54) similarly acknowledges that concern for good governance in a development context is both 'appropriate and useful,' as it draws attention to the previously neglected 'political and administrative aspects of development.' Good governance is an inherently political concept. One problem with the prevailing version of good governance is precisely its purportedly apolitical, technocratic approach. This is, in fact, politically charged, because it dismisses the participatory elements essential for sustainable development.

As Meadowcroft (2007, 313) underlines, governance for sustainable development 'is not something that competent officials can get done quietly and efficiently, out of the public view,' but 'requires societal decisions about desirable ways of life, and about how benefits and burdens are to be shared among different communities and different generations, and between humankind and other inhabitants of this planet.'

A meaningful contribution to the good governance debate would therefore involve moving beyond the narrow, structural-adjustment-driven conception of good governance promoted by global institutions, and thinking instead more broadly about 'what sort of world we want to inhabit, and how we can coordinate getting there' (Evans, 2012, 14). A crucial first step in this process is to critically assess the limitations of existing good governance models and the practical challenges of applying them – despite their purported simplicity and broad applicability. Beyond this, there is a need to return to the original spirit of the term and reclaim good governance as a radical, transformative idea. But the answer to what makes governance good will not be a singular, universal one, but rather 'socially constructed and culturally bound' (Massey, 2022, 81). As Massey (2022, 82) underscores:

> It is impossible to divorce decision-making in the real world from the context of history, struggle, ethnicity, wealth and power. It is perhaps more fruitful to explore ways of confronting these realities and making them work in favour of good governance than to adopt some abstract notion that pretends they do not exist.

For these reasons, we agree with Andrews (2008, 399) that 'it is unlikely that such a framework will be neat and tidy like the current good governance literature appears to be.' In this context of increased complexity, it may also be helpful to consider Grindle's (2012) concept of 'good enough governance.' This recognises that 'not all governance deficits need to be (or can be) tackled at once' (Grindle, 2012, 268). Some governance objectives may be more or less relevant or urgent depending on the specific governance context. Governance priorities must be set and 'assessed in light of historical evidence, sequence, and timing' (Grindle, 2012, 268), and may need to be flexible through time.

Overview of the volume

This book is divided into two parts, concerned with the overarching questions of 'what makes governance good?' and 'what are the obstacles to good governance at present?,' respectively. Part I, *Good governance: realities and possibilities,* explores the current state of governance and emerging definitions of good governance across different policy areas. Part II, *Challenges to good governance*, examines contemporary obstacles to good governance that cut across policy areas, along with broader questions about the relationship between governance and development. Collectively, the chapters in this volume also raise and address related subquestions, such as:

- What is the current state of governance across the world, and across different policy areas?
- How do current realities align with the ideal of good governance?
- What changes are needed to achieve good governance and sustainable development?
- Can the challenges to good governance be overcome?

This book thus explores the multidimensional (local, national, supranational, global) and interdisciplinary nature of governance, drawing on case studies from diverse contexts and policy areas, while also showing how governance levels and policy areas intersect. We start from an open ontology – the question of what good governance for sustainable development might look like – which means that we are necessarily also methodologically open. Recognising that good governance is ultimately context-specific, we intentionally avoid restricting methodology or prescribing a set methodological approach. Nevertheless, each chapter will, to some extent, engage with both empirical evidence and with the theoretical-normative dimension implied by the adjective 'good' in 'good governance.'

Part I *Good governance: realities and possibilities* explores the question of what constitutes good governance across different policy areas that broadly correspond to the three pillars of sustainable development (environmental, economic, and social), across Chapters 2, 3, and 4. Despite their different emphases, the three pillars intersect and interact, and all require a reckoning with colonial legacies, as Chapter 5 suggests, in framing good governance as inherently decolonial.

Part I begins with a chapter on environmental governance. In Chapter 2, Jessica Savage explores the disconnect between global environmental policy and the needs and practices of local communities in their interactions with the environment. The chapter traces the historical development of environmental governance, noting that environmental policy – as we now recognise it – emerged relatively late on the global stage, following the growing awareness that human activities were having a profound impact on ecological systems. A key argument is that humans have long interacted with the environment, well before the advent of formal environmental governance, which arose not purely out of concern for nature, but as a response to the excesses of capitalist development. In this sense, environmental governance is framed as a mechanism to moderate humanity's most harmful impacts on the environment, rather than being about the environment *per se*. Within this context, a provocative question is posed: might good governance, in some instances, actually mean no governance at all? The environment, the chapter suggests, is inherently self-regulating and did not require management until it was subordinated to human exploitation. Contemporary environmental policy is thus caught in a fundamental contradiction: on one hand, advocating for integration and centralisation of governance, and on the other, calling for disintegration, decentralisation, and localisation. It is too late to not manage the environment. But the pressing question remains: how can we govern it effectively while mitigating the most damaging effects of capitalism? The final sections of the chapter address this question with a focus on successes and failures of global environmental policy to date and the lessons that can be drawn from these.

Chapter 3, by Maria Gavris and Jason Heyes, develops a theory of good labour governance grounded in degrowth principles and informed by the current limitations of global labour governance. The chapter opens with an overview of the current state of labour governance globally, highlighting persistent challenges despite decades of reform, which range from limited participation in and coverage of labour governance, to weak enforcement of labour standards, and fragmented labour processes benefitting large corporations. Until recently, labour policies have also largely been conducted in a siloed approach detached from ecological concerns, but this has changed in the context of recent crises, with green growth and degrowth entering the debate. The chapter examines these perspectives, with a particular focus on the latter, to explore what role labour might play in a sustainable world. It emphasises the need for both quantitative and qualitative changes in the world of work: reductions in working time to match a shift away from the profit-oriented market economy in the context of a broader redefinition of work, its purpose, and its societal and environmental value. It then examines implications for labour governance, including ways to address some of the present issues, as well as controversies around the role of the state in supporting a sustainable transition, and the pressing need to tackle existing global power relations and strengthen multilevel governance. Finally, the chapter also considers remaining issues in the application of degrowth principles to labour governance, as identified from the labour movement's incipient engagement with degrowth ideas. Ultimately, based on this analysis, the chapter argues that good labour governance must consider both the value of work to society as well as the experiences and identities of those performing the work.

Marta Guerriero focuses on the relationship between good governance and gender equality in Chapter 4. Traditionally, gender inequalities were assumed to arise exclusively within the private sphere, but more recently inequality has increasingly come to be understood as multidimensional, with significant public and governance dimensions. Yet the mainstream good governance agenda remains largely inattentive to gender and its intersections with race, class, ethnicity, sexuality, and to the ways in which these intersections influence who participates in governance and how the outcomes of governance are experienced by different people. In practice, national and international organisations do incorporate gender considerations into policy design – partly in response to the global Sustainable Development Agenda – but these efforts are often fragmented and have limited impact on gender inequality. One reason for this limited impact is women's continued exclusion from decision-making and political life: they are frequently treated as vulnerable groups rather than as agents of change. Overall, the chapter argues that there is a dialectical relationship between gender equality and good governance: good governance promotes gender equality in politics, and gender equality is necessary for genuinely good governance. The chapter concludes that good governance requires inclusivity and accountability, but these concepts become hollow if intersectionality is ignored. After all, how can policies be considered accountable and inclusive when many of the voices to whom they should be accountable are excluded from the conversation?

In Chapter 5, Vanessa Burns explores decolonial governance as good governance, arguing that there is a moral, prudential, and pragmatic imperative to decolonise existing governance structures. The chapter begins by unpacking various interpretations of decolonisation, from radical to reformist, and from backward-looking reparative approaches to forward-looking transformative ones. These perspectives are then applied to environmental governance, drawing on the author's expertise in climate change adaptation policy for small island states. In this context, good environmental governance is defined as one grounded in indigenous knowledge, challenging the Western epistemic divide between humans and nature. However, the chapter then shows how such a mode of governance continues to be challenged by international institutions and norms, and specifically the colonial trope of 'lack of capacity,' used in policymaking circles to dismiss indigenous knowledge and concerns. The chapter then turns to concrete examples, to illustrate both the challenges of decolonising environmental governance in small island states, focusing on the Coral Triangle, and successful decolonial practices. The examples of women's custodianship of spring systems in Timor-Leste and iTaukei hybrid governance in Fiji are used to illustrate successful indigenous governance systems that challenge colonial-inherited models. The chapter closes with a reflection on good governance itself, arguing that it requires commitment to deep institutional change, meaningful engagement with epistemic alternatives, and a supracolonial approach that confronts the shadow of colonialism while recognising that the history of colonialism cannot be simply erased.

Part II, *Challenges to good governance,* begins with a chapter on the challenges of translating macro-level sustainable development initiatives into sustainability actions on a micro level. In Chapter 6, Molly Wimonmat Srichamroen uses Thailand as a case study, drawing on semi-structured expert interviews to compare sustainability efforts and governance practices across the public and private sectors. The chapter reveals that even the mainstream good governance agenda, praised for its alleged simplicity and universality, is actually very difficult to implement in practice, not least because it appears to be more directly applicable to state-level action than to levels below the state. Although Thailand's national sustainable development policy explicitly incorporates good governance criteria, it remains unclear to what extent these are understood, adopted, accepted, and operationalised below state level, within public and private organisations. The chapter finds significant divergence: private sector organisations often equate sustainability with business continuity, with governance driven by stakeholder expectations and growth goals; public sector bodies, while embracing a broader mandate focused on public welfare and quality of life, often fail to address internal governance challenges such as overwork, environmental harm, or institutional corruption. These findings highlight the need to focus not only on governance outcomes but also on governance processes, and how these processes affect outcomes, in judging the goodness of governance in a particular setting.

Complementing this, Chapter 7, by Anastasia Ershova, explores the challenges of applying good governance principles in supranational contexts. Focusing on the European Union (EU) as the most advanced system of supranational governance in the world, used as a model for other regional integration projects globally, the chapter assesses the EU's own good governance framework, developed in response to recurring legitimacy crises and perceptions of a democratic deficit. Specifically, it centres on participation as a principle of good governance, and its relationship to responsiveness and input legitimacy. The chapter finds that even the European Commission, the EU's most technocratic institution, occasionally aligns with public sentiment, but cautions that its actions may still mainly reflect institutional self-interest. Moreover, even when the Commission's behaviour reflects average public opinion, it may mask significant differences both between and within member states. As such, it remains unclear which public demands, and why, carry influence over European-level policymaking. This undermines the participation principle and casts doubt over the EU's ability to adhere to its own principles of good governance, raising bigger questions about the sustainability of supranational systems. Together, Chapters 6 and 7 thus challenge the presumed universality and internal coherence of the existing good governance agenda – even on its own terms. If the supposed strengths of this model (its simplicity and ability to offer a neat, comparable framework across contexts and over time) fail to materialise in practice, then the case for moving beyond it towards more critical good governance approaches becomes even more compelling. At the same time, these chapters underscore that the persistent difficulties of multilevel governance must be addressed, regardless of how good governance is defined.

Chapter 8, by Dan Coffey and Carole Thornley, identifies existing global dependencies and power relations as key obstacles to achieving good governance today. It focuses on the car industry – a significant contributor to global warming and unsustainability – as a case in point. After reviewing two waves of car criticism, this chapter outlines three governance perspectives on the industry: naïve, balanced, and bleak. The naïve view is characterised by technological optimism, assuming that innovation alone can ensure sustainability, ignoring how resources are sourced and overlooking the social and environmental costs of resource extraction. The balanced view takes a broader approach, accounting for community rights, ecological impacts, and resource conflicts. The bleak view goes further, arguing that as long as the profit motive and extractivist logic dominate, the concerns of the balanced view cannot be meaningfully addressed, rendering good governance unattainable. Part of the problem is that the car is, for many, an 'aspiration,' synonymous with 'development.' In this sense, the car industry is both ungoverned and effectively ungovernable, because dominant interpretations of development shield it from critical scrutiny. Technological fixes are no solutions at all, because while they claim to improve 'sustainability,' they do not attempt to alter the status quo on 'development' in any way. Therefore, this chapter illustrates that pursuing genuinely good governance requires rethinking the very meaning of development – an immensely difficult task, given prevailing power dynamics.

Finally, Chapter 9, by Maria Gavris and Jessica Savage, concludes this book by synthesising its diverse transdisciplinary contributions and reflecting on the overarching questions introduced above, around what constitutes good governance for sustainable development, and what challenges stand in the way of achieving it. The chapter formulates a definition of good governance as decolonial, interdisciplinary, inclusive, intersectional, democratic, and requiring action at multiple levels, but also context-dependent, multifaceted, historically informed, and socially constructed. A central theme reiterated here is the contested and value-laden nature of both 'good governance' and 'sustainable development.' In this light, fundamental questions around how progress is defined, and who defines it, pose significant obstacles to governance that is genuinely sustainable. Global power relations and dependencies are discussed as significant challenges to good governance. The chapter also identifies key areas for future research: how to meaningfully decolonise governance (including uncovering what has been erased in cementing the current good governance agenda), how to develop good governance methodologies aligned with an open ontology, and whether the ongoing global polycrisis is prompting a critical re-evaluation of mainstream governance approaches. Overall, the concluding discussion argues that governance for sustainable development needs to move beyond a superficial exercise, and must instead interrogate the deeper assumptions underpinning dominant development models.

References

Adejumo-Ayibiowu, D. 2015 'A theoretical basis for good governance,' *Africanus: Journal of Development Studies*, vol. 45(2), pp. 1–27.

Adelman, S. 2018 'The sustainable development goals, anthropocentrism and neoliberalism,' in French, D. and Kotzé, L. (eds.) *Sustainable Development Goals: Law, Theory and Implementation*. Cheltenham: Edward Elgar, pp. 15–40.

Andrews, M. 2008 'The good governance agenda: beyond indicators without theory,' *Oxford Development Studies*, vol. 36(4), pp. 379–407.

Brundtland, G. 1987 Report of the World Commission on Environment and Development: Our Common Future. United Nations General Assembly Document A/42/427.

Costanza, R. and Patten, B. 1995 'Defining and predicting sustainability,' *Ecological Economics*, vol. 15, pp. 193–196.

Esquivel, V. 2016 'Power and the sustainable development goals: a feminist analysis,' *Gender & Development*, vol. 24(1), pp. 9–23.

Evans, J. P. 2012 *Environmental Governance*. London: Routledge.

Gisselquist, R. M. 2012 'Good governance as a concept, and why this matters for development policy,' WIDER Working Paper, No. 2012/30, The United Nations University World Institute for Development Economics Research (UNU-WIDER), Helsinki.

Grindle, M. 2012 'Good governance: the inflation of an idea,' in Sanyal, B. et al. (eds.) *Planning Ideas That Matter: Livability, Territoriality, Governance, and Reflective Practice*. Cambridge: MIT Press, pp. 259–282.

Gruffydd Jones, B. 2014 ''Good governance' and 'state failure': the pseudo-science of statesmen in our times,' in Anievas, A., Manchanda, N. and Shilliam, R. (eds.) *Race and Racism in International Relations*. London: Routledge, pp. 62–80.

Haifa, A. A., Zineb, O. T. and Ziad, Q. 2023 'Good governance: concept, basics, and relationship to sustainable development from a Palestinian perspective,' *Sustainable Development*, vol. 31(4), pp. 2122–2136.

Icaza, R. 2022 'Decoloniality, governance and development,' in Hout, W. and Hutchison, J. (eds.) *Handbook on Governance and Development*. Cheltenham: Edward Elgar, pp. 45–62.

Kanie, N., Zondervan, R. and Stevens, C. 2014 Ideas on Governance 'of' and 'for' Sustainable Development Goals: UNU-IAS/POST2015 Conference Report. Tokyo: United Nations University Institute for the Advanced Study of Sustainability.

Kooiman, J. 2003 *Governing as Governance*. London: Sage.

Kopnina, H. 2016 'The victims of unsustainability: a challenge to sustainable development goals,' *International Journal of Sustainable Development & World Ecology*, vol. 23(2), pp. 113–121.

Kotzé, L. 2018 'The sustainable development goals: an existential critique alongside three new-millennial analytical paradigms,' in French, D. and Kotzé, L. (eds.) *Sustainable Development Goals: Law, Theory and Implementation*. Cheltenham: Edward Elgar, pp. 41–65.

Kuhlman, T. and Farrington, J. 2010 'What is sustainability?' *Sustainability*, vol. 2, pp. 3436–3448.

Massey, A. 2022 'Sustainable development goals and their fit with good governance,' *Global Policy*, Vol. 13(Suppl. 1), pp. 79–85.

Meadowcroft, J. 2007 'Who is in charge here? Governance for sustainable development in a complex world,' *Journal of Environmental Policy & Planning*, vol. 9(3–4), pp. 299–314.

Mkandawire, T. 2007 'Good governance: the itinerary of an idea,' *Development in Practice*, vol. 17(4–5), pp. 679–681.

Plumptre, T. and Graham, J. 1999 'Governance and good governance: international and aboriginal perspectives,' Institute on Governance publication, Available from https://www.files.ethz.ch/isn/122184/govgoodgov.pdf (accessed 12 February 2025)

Purvis, B., Mao, Y. and Robinson, D. 2019 'Three pillars of sustainability: in search of conceptual origins,' *Sustainability Science*, vol. 14, pp. 681–695.

Strange, S. 1996 *The Retreat of the State: The Diffusion of Power in the World Economy*. Cambridge: Cambridge University Press.

Streeten, P. 1992 'Global governance for human development,' UNDP-HDRO Occasional Papers No. 1992/4. Available from https://ssrn.com/abstract=2294579 (accessed 12 February 2025)

Tooze, A. 2021 *Shutdown: How Covid Shook the World's Economy*. London: Allen Lane.

UN 1998 Annual Report of the Secretary-General on the Work of the Organization. Available from https://digitallibrary.un.org/record/259420?ln=en&v=pdf (accessed 27 January 2025).

UN, 2009 What is good governance? Available from https://www.unescap.org/resources/what-good-governance (accessed 10 February 2025).

UNDP, 2011 *Towards Human Resilience: Sustaining MDG Progress in an Age of Economic Uncertainty Report*. New York: United Nations Development Programme Bureau for Development Policy.

van Arkadie, B. 2012 'Good governance and donors,' in Sundaram, J. K. and Chowdhury, A. (eds.) *Is Good Governance Good for Development?* London: Bloomsbury, pp. 53–70.

White, M. 2013 'Sustainability: I know it when I see it,' *Ecological Economics*, vol. 86, pp. 213–217.

World Bank 1989 *Sub-Saharan Africa. From Crisis to Sustainable Growth: A Long Term Perspective Study*. Washington, DC: World Bank.

World Bank 1992 *Governance and Development*. Washington, DC: World Bank.

World Bank 2024 Worldwide governance indicators, 2024 Update, World Bank https://www.worldbank.org/en/publication/worldwide-governance-indicators (accessed 10 February 2025)

PART I

Good governance

Realities and possibilities

2

RE-GOVERNING THE SELF-GOVERNING

Environmental governance is not about the environment

Jessica Savage

Introduction

Normative discussions surrounding notions of environmental conservation and restoration focus on 'saving the planet.' The uncomfortable truth here is that, if humans were to disappear from the planet tomorrow, the Earth would be just fine. It may take time to recover, but our ecosystems would find a new equilibrium. The issue is humans, or rather a growth-focused society with demands exceeding the resources our environment is able to provide. Critically, this means that governing our environment is not really about our environment at all, but instead about altering the way that humans – individuals and societies – interact with the environment.

We are living in a period of unprecedented environmental change, most recognisably through the somewhat vague normative concept of 'climate change' but there is a series of interconnected simultaneous crises evolving, relating to biodiversity loss, ocean acidification, water accessibility, waste management, etc. Unfortunately, the list is long, but what all of these issues have in common is that they evolved as a result of our societies' actions, and the associated impacts on environmental equilibrium. There have been numerous times in the Earth's history where the planet has undergone significant periods of environmental change (Rockström et al., 2009). Historically, these changes have occurred as a result of natural perturbations in the ecological balance of the planet. The difference now is that human actions have triggered changes to occur at a rate much faster than seen before. The environmental policy landscape is fast-evolving and complex, and has emerged in an attempt to manage declining environmental health. This chapter explores the complex relationship between people, policy, and planet to explore ways

DOI: 10.4324/9781003467694-3

to govern and re-govern the self-governing. Real environmental change will come from managing the interactions between policy and people, but that is not as simple as it sounds.

This chapter will first provide an explanation of environmental governance, particularly in the context of the interactions between humans and the essential natural resources provided by their environment. It will highlight that historically the environment has been a self-governing entity, but this changed with industrialisation and humanity's obsession with growth. This chapter will then provide an overview of the successes and failures of supranational environmental policy. Using the example of the Convention on Biological Diversity (CBD), and other key supranational policies, this chapter makes recommendations for steps towards a model of good environmental governance, highlighting the need to go back to basics, and ensure that environmental targets fully integrate the human dimension, in an achievable and step-by-step manner.

Governing natural resources

Governance can generally be defined as 'the interactions among structures, processes, and traditions that determine how power and responsibilities are exercised, how decisions are taken, and how citizens or other stakeholders have their say' (Graham et al., 2003, as cited by Lockwood et al., 2010, 987). However, there is growing recognition, at least within circles concerned with the governance and management of natural resources, that there is a need for greater levels of integration and coordination (Lockwood et al., 2010). New governance arrangements should focus more on the multiscalar (spatial and temporal) phenomena characteristic of environmental systems (Lemos & Agrawal, 2006). The concept of *new governance* (Howlett & Rayner, 2006) is often used to describe governance which prioritises collaborative approaches among government and non-government actors from both the private sector and civil society (Lockwood et al., 2010), and while this concept may now not be considered particularly *new* it does represent a valuable step towards good governance in terms of managing our environment.

Key changes led to the need to re-specify the definitions of governance in this environmental context. Primarily these link to the recognition of the polycrisis we collectively find ourselves facing, and recognising the need for responses to environmental 'wicked problems.' Recognising the interdependencies within ecological rules and relationships, compounded by the concurrent interdependencies between humans and their environment, governance mechanisms need to have awareness of these interdependencies when coordinating the actions and decisions of diverse actors (Lockwood et al., 2010). Put more simply, there is a recognition in environmental governance that global challenges require global responses.

However, perhaps ironically, we now find ourselves in the contradictory situation where normative discourse leans towards integration and centralisation of governance in one instance, and then simultaneously, there is a second discourse prioritising disintegration, decentralisation, and localisation (Rosenau, 2000). Understanding the complexity of these conflicting narratives is critical in the context of how we respond to environmental degradation, tying specifically into the ways that supranational policy mechanisms designed to respond to globalised wicked problems result in shifts in both authority and power upwards from national to supranational scales through the application of international conventions, but also downwards to subnational and local scales through the devolution of central government responsibility (Pierre, 2000; Lockwood et al., 2010).

This issue of scale is critical, particularly given the global polycrisis we find ourselves facing. However, in some ways the situation is more complex because historically there was less of a need for environmental policy, in part due to human activities only exerting excess pressure on environmental function relatively recently, but also in part due to the naturally occurring resilience in environmental systems through ecological equilibria.

Self-governing systems

Ecological communities are structured by a unique balance between the species present, and the interactions both within and between those species, and their environment (Singer, 2016).

These species and interactions exist and operate across all temporal and spatial scales, resulting in a planet which functions through a complex balancing act constructed of immeasurable, tiny, and seemingly inconsequential actions and reactions within the biotic (living) and abiotic (non-living) environment combining to provide a sustainable equilibrium. That equilibrium has existed for millennia, providing a *moderately* stable state from which species and ecosystems have evolved, and are continuously evolving.

The keyword here is moderately. In such dynamic systems a stable equilibrium does not mean static; in fact, the flexibility and adaptability of our ecosystems, due in part to their dynamic nature, have provided a good degree of resilience and recovery. Unfortunately, we are now reaching the point where the external pressures and disturbances are so great, that naturally occurring flexibility and resilience are no longer enough to continue to provide the stable equilibrium we need (Rockström et al., 2009). Why does this matter? Because, like it or not, we are solely dependent on the function of our ecosystems and broader biosphere to provide everything we need to survive.

The way our planet functions is based on a complex combination of ecological rules connecting individuals within and between species with their environment. However, a certain simplicity comes from the fact that those ecological rules are the same across all scales, from petri-dish to planet, ideally working together to find a stable equilibrium. Ecosystem stability is essential, and is a well-researched topic within ecology, with a great deal of focus on how properties of disturbance affect ecosystem response (Zelnik et al., 2018). However, the spatial and temporal dimensions of both the ecosystems and associated disturbances add more layers of complexity which, at least in part, explains why understanding these dynamics on a global scale has proven extremely challenging (Zelnik et al., 2018).

There is a natural limit to the number of individuals of a given species that their environment can support. If we look at this through a human lens, there is a limit to the number of humans that the planet Earth can support, particularly if we consider quality of life as a desirable factor (Siegel, 2021). This limit is decided by factors such as the availability of natural resources (water, food, space) and the management of waste products (i.e. pollution). However, the same moderately stable conditions that allowed many species to evolve and develop allowed humans to do the same. What is the difference here? Humans are the first species (that we know of) who have both the intelligence to develop industrialised societies, and that lack the wisdom to do it properly. We have let development destabilise the environmental equilibrium that both facilitated and sustained it. Development impacts environmental function in a myriad of ways: for example, deforestation to make space for the development of agriculture and urban areas reduces the space for plants and animals to grow and forage. Perhaps more critically, this disruption also alters biogeochemical and nutrient flows, impacting ecosystem function (Rockström et al., 2009). Without that ecological equilibrium, we run the risk that the environment will no longer be able to continue to provide the ecosystem services that we depend on (Rockström et al., 2009; Myers et al., 2025), which leaves us with the responsibility to manage ecosystem function ourselves. We often hear the notion that we need to 'save the planet,' but the reality is that we need to maintain the planet in a state that is able to support humans.

Environmental policy to re-govern

People have been living with the environment for a long time, much longer than we have had environmental policy. It was only when the human population began to increase towards the carrying capacity of the Earth that we started to detect the negative impacts of human activities which necessitated the development of policies as a response mechanism. Understanding the exact carrying capacity of the Earth for humans has been debated for centuries, long before many issues emerged. In 1679, Antoni van Leeuwenhoek estimated that the maximum carrying capacity would be 13.4 billion – calculated by scaling the estimated population of the Netherlands by the estimated inhabitable space on earth, crude but reasonable (van Leeuwenhoek, 1679;

Cohen, 1995). More recent estimates suggest a range from 3 to 44 billion (Cohen, 1995), a broad range admittedly, but at least allowing for some consideration of human comfort. Understandably, these numbers should be viewed with a degree of scepticism; however, the very nature of that scepticism allows us to continue to delay the inevitable need to address issues of human-induced environmental change.

Technological advancement bought us more time, allowing us to artificially manage the changing environment, enhancing food production through agricultural improvements, and water access through desalination and filtration methods (Siegel, 2021), but it has not been enough. However, there are two key components in the equation which calculates, or at least estimates, the Earth's carrying capacity for humans: the first is the 'natural resources' factor, discussed in the ecological definition of carrying capacity. But the second, perhaps more pivotal factor in this equation relates to human behaviour, and human choices in terms of economics, culture, and use of the environment (Cohen, 1995). Choices and behaviour suggest an individualistic approach, which while valid, fails to recognise the true scale of the current, and growing, population, and the scale at which interdependency needs to be addressed within governance. This has naturally led to the development of environmental policy as the 'natural' next step. But it is important to understand that environmental policy has almost nothing to do with managing the environment, and everything to do with managing people and their interactions with the environment.

Numerous disciplines examine human-ecosystem interactions, highlighting factors where environmental outcomes influence human wellbeing (Cox et al., 2016) and vice versa. But these approaches keep humans on the periphery of nature, without recognising that we are a part of nature. Normative Global North discourse suggests that we can dominate and master nature like we did people in the Global South. But critically, this is not about us being the masters of nature, we have to recognise that we are, first and foremost, a part of nature. Most Global North methods in the way we manage nature mirror the ways in which colonialism managed people. Domination, extraction, ownership. We have done the same thing to nature because nature does not have a voice. It took until environmental stability was no longer guaranteed, before we recognised there was a problem.

A brief(ish) history of environmental policy

Environmental policy, in the guise we would recognise, emerged comparatively late on the global policy stage, tied to the relatively recent realisation that human actions were impacting environmental function globally. Critically, this happened once we recognised that such impacts are also present in a way that actions in one part of the world may not impact the surrounding area, but manifest on the other side of the planet.

While there are numerous earlier examples of regional, national, and local policies designed to improve the environment, it was only in 1971 when the Ramsar Convention on Wetlands emerged to deal with the declining state of wetland habitats, that countries truly began to collaborate towards a collective response to environmental decline (Gardner & Davidson, 2011). The Ramsar Convention is generally considered to be successful, due to the number of signatory countries and the number and global spread of Ramsar sites leading to a reduction in the rate of ecological degradation (Rattan et al., 2021). However, the reach of the Ramsar Convention was always quite small, because despite the global focus, only one habitat type was targeted.

Later, in 1983, when the World Commission on Environment and Development produced the report 'Our Common Future' (commonly referred to as the Brundtland Report), environmental issues were more recognised on the global political stage (Brundtland, 1987), particularly given the broad environmental focus. The Brundtland report is perhaps best known for outlining the concept of sustainable development. Critically, the report not only described sustainable development but also explored key causes of environmental degradation, and made novel attempts to understand the complex interconnectedness of social (in)equity, our economy, and the environment. The legacy of the Brundtland report is still evident in the normative discourse in environmental policy, although perhaps most pivotally, the report outlined the working definition of Sustainable Development ('development that meets the needs of the present without compromising the ability of future generations to meet their own needs') (Brundtland, 1987). However, while this was a pivotal moment in the formation of environmental governance, at least as we recognise it today, it is clear that almost 40 years later, we really have not made much progress.

In the 1990s, the UN sponsored a series of world summits, such as the UN Conference on Environment and Development (UNCED, also known as the Rio Earth Summit), providing an innovative new platform for global governance. This offered the opportunity for the largest collection of government parties in human history, and critically concluded in the establishment of two key treaties designed to address key global environmental challenges: the CBD and the UN Framework Convention on Climate Change (UNFCCC) (McGraw, 2002), more commonly known as the Kyoto Protocol and later the Paris Accord due to subsequent agreements during conference of parties proceedings in the relevant cities.

As we came into the new millennium, supranational environmental policies started to become more visible, perhaps merely due to their somewhat rapid implementation. Notably the Millennium Development Goals (MDGs) were eight global targets established by the United Nations, with aims towards reducing poverty, gender inequality, and disease, while simultaneously boosting environmental sustainability and political cooperation (Sachs, 2012).

However, progress was uneven, and despite efforts, the MDGs failed to meet the bold targets set within their allotted 2000–2015 window (Fehling et al., 2013). Reviews attributed issues to both flaws in the design of the framework, and extrinsic issues (Fehling et al., 2013). Subsequent targets emerged in 2015 in the form of the Sustainable Development Goals (SDGs). Continuing from their predecessors, these 17 goals have similar objectives as the MDGs, although they have been expanded to include 169 targets with the aim to 'inspire and guide policies for eradicating poverty, protecting planetary ecosystems, and promoting peace and prosperity for humankind' (Ma et al., 2025; United Nations, 2015). Designed to operate between 2015 and 2030, these targets appear to be falling into the same trap of uneven implementation as their predecessors, with analysis suggesting that different countries are performing differently in targets, and pursuing alternative paths towards sustainable development (Ma et al., 2025). This puts us in a difficult position, where we must simultaneously acknowledge that a 'one-size-fits-all' approach is rarely appropriate given the so-called 'wicked problems' we collectively face, while also acknowledging that if we are all doing things in our own way, we might not actually be making any progress towards common goals.

International politics around multiple centres of power have become the model of contemporary international policy. However, these policies are often characterised by rising uncertainty, interdependence, multiple actors, competing legitimacies, complex transnational relations, all further complicated by the nature of the problems and responses to them (Le Prestre, 2002). Notably, those supranational policies which evolved from conventions linked to single habitats or issues have historically been more successful (i.e. Ramsar and Montreal), whereas the broader conventions trying to deal with multiple elements of the environmental polycrisis typically struggle. Yet often, the struggle is not acknowledged. Is this to save face? By consistently shifting targets, do we build momentum, or do more harm than good?

What counts as success in environmental policy

While not necessarily surprising, the response to a lack of success in policy tends to prioritise optics over progress. So, what we end up with is an increasing number of supranational policies aimed at setting targets which are unrealistic. A good example here is the CBD. First adopted at the Rio Earth Summit, the CBD is a broad supranational policy framework with the purpose to maintain the biological diversity of our planet, in order to prioritise ecosystem function and support the ecosystem services provided by our planet and used by its inhabitants. At the time of its adoption, the CBD was considered a progressive development within international law, introducing the concept that biodiversity is a *common concern for all human kind* (Francioni, 2014).

While the semi-prescriptive nature of the CBD can make things complex, at its heart, the intention was for the CBD to act as a mechanism to encourage/require the implementation of protected areas as a means of conserving biodiversity (Agardy et al., 2003). This stems from a logical scientific foundation based on the principle of carrying capacity. A certain amount of space has a limit for the number of individuals it can support. If you minimise human disturbance in an area, and populations are allowed to recover, after a period of time populations should theoretically increase, at which time those individuals leave the protected area in search of more natural resources they need for survival. This is known as the 'spill over effect' (Cumming, 2025), and is a fundamental principle in conservation ecology. The principle of a protected area has been around for about as long as humans have been capable of building fences, and the science behind their function is quite simple, at least on land, where it is easy to count species and monitor improvement. Things get more complicated in aquatic environments, because it is not realistic to put a fence in the ocean, and monitoring the change as a result of the protection becomes both more complicated and more expensive (Savage et al., 2020). Aside from the practical, the rate of recovery will naturally vary, depending on the species you are intending to conserve or save.

The CBD provides a flexible framework, designed to accommodate developed and developing countries' concerns and capacities (Morgera et al., 2014), at least on paper. However, implementation across geopolitical scales is very challenging, because biodiversity decline is a worldwide problem (experienced by all countries, but resolved in a way that is largely national) rather than a 'global' problem (affecting all countries indiscriminately and resolved on a largely global scale) (Le Prestre, 2002). In this context, critics also highlight that the traditional Western-model of international law encourages states to privatise natural resources under the guise of biodiversity protection, in opposition to the more progressive view that natural resources should be considered common pool resources (Francioni, 2014).

The CBD has three main areas of operation:

1. Protecting biodiversity
2. The sustainable use of its components
3. The fair and equitable sharing of the benefits generated by the exploitation of genetic resources.

The latter of these targets has resulted in the following sub-protocols, providing more specified frameworks for knowledge sharing.

- *The Cartagena Protocol on Biosafety* – Focuses on the safe handling and transport of living modified organisms resulting from biotechnology, that may have adverse effects on biological diversity, and pose a risk to human health (Pythoud & Thomas, 2002).

- *The Nagoya Protocol on Access to Genetic Resources and the Fair and Equitable Sharing of benefits arising from their utilisation* – Provides a transparent legal framework for the effective sharing of knowledge relating to the use of genetic resources (Blais, 2002).

However, points 1 and 2 above are primarily managed through the implementation of protected areas. But the mechanisms of implementation are as contentious as they are complex. In 2002, the 193 signatory parties to the CBD agreed to 'significantly reduce the rate of biodiversity loss by 2010.' This commitment, known as the 2010 biodiversity target (Jones et al., 2011) was later expanded to include additional sub-targets for the establishment and maintenance of terrestrial protected areas by 2010 and marine protected areas by 2012 (Savage et al., 2020), requiring these protected areas to be 'comprehensive, effectively managed and ecologically representative of national and regional ecosystems' (Toropova et al., 2010, 8). Further investigation in the manifestation of these targets highlights an interesting pattern:

- 10% of the land and sea in protected areas by 2010/2012 – CBD (Rodriguez-Rodriguez et al., 2011; Toropova et al., 2010)
- 20% of the land and sea in protected areas by 2020 – Aichi Target 11 (Visconti et al., 2019)
- 30% of the land and sea in protected areas by 2030 – *World Parks Congress* (Schugart-Schmidt et al., 2015)

While some may perceive increasing targets as progressive, when reviewed in partnership with data tracking on progress towards these targets, the picture becomes much less optimistic. In October 2025, Protected Planet, the organisation responsible for collating global data on marine and terrestrial protected areas reported 16.45% of terrestrial and inland waters (n = 287,225) and 9.61% (n = 16,545) of the marine environment were within protected areas. These numbers improve slightly to 17.55% and 9.85% respectively, when broadening definitions to include sites described as providing 'Other Effective Conservation Measures.' This is still a far cry from CBD targets, and monitoring the data presented by Protected Planet over time shows irregular fluctuations in the data, as many protected areas are declared, but also downgraded. Furthermore, species population and abundance outside of protected areas continue to decline (Visconti et al., 2019).

Part of the issue here is that while the targets for the declaration of protected areas are clear, guidance on their success just suggests that they should be both effective and representative of all (Agardy et al., 2016), but the reality is much more complex. Given the devolved nature of the way that protected areas are declared from the supranational bodies dictating the need, and the vast array of local contexts that must be considered, it would be impractical to suggest that a 'one-size-fits-all' approach to implementation, monitoring, and management would be appropriate. This is not forgetting the practicalities of protected areas, where on land they work through the use of signs and fences, creating physical barriers between the protected environment and the humans they are protected from. The unfortunate reality is that you cannot put fences in the oceans, and monitoring marine protected areas is particularly tricky given that humans generally struggle to breathe underwater without assistance.

The elephant in the room is that we really do not have an alternative strategy. The science of a protected area is sound. If you want to protect an ecosystem/habitat from harm, keep the plants and animals in, keep the humans out. This is very logical albeit not always practical. It is based on methods of keeping livestock from being stolen, or wandering off. But that is often insufficient, because even where it is possible to put up a fence, poachers still break in. And where it is not possible to put up a fence, such as in the sea, one could forgive fishers for losing their way and fishing inside a protected area. Radical steps have been taken to try to address these issues including personal armed guards for critically endangered species (Barichievey et al., 2017), and GPS trackers for fishing vessels that monitor movements (Natsir et al., 2019). While armed guards have not proved particularly effective (Barichievey et al., 2017), technological responses do show more promise (Natsir et al., 2019; White et al., 2020).

Unfortunately, this leaves us in a difficult situation. We are constantly increasing unmet targets in a bid to maintain momentum, but lack appropriate mechanisms to support the fulfillment of said targets. This leads to a large implementation gap, resulting in a vast quantity of what is known in conservation circles as 'paper parks,' with estimates suggesting that up to 27% of marine protected areas may be failing to meet their management objectives (Relano & Pauly, 2023), although that number is realistically much higher. Which begs the question, are we prioritising failure, or is it enough to hope for change?

A policy regime can be considered effective when it makes a notable contribution towards improving the issue it was created to address (Le Prestre, 2002). However, others equate effectiveness with compliance, but critically different disciplines often view success through different lenses: lawyers emphasise implementation and compliance, scientists and environmentalists tend to be more focused on problem-solving and meeting goals, philosophers prioritise normative aspects, while political scientists identify behavioural change and cooperation as priorities (Le Prestre, 2002).

Adding to this complexity, global policy has seen a shift in recent decades towards devolution, decentralisation, downsizing, and supposed debureaucratisation, combined with privatisation, contracting out, and the integration of business management techniques, there is now wider recognition that much of the work is not done by government organisations, instead depending on partnerships and broader public and private networks and organisations (Roberston & Choi, 2010). Critically, this generates implementation difficulties, especially in communities that are resource-limited. The costs, logistical constraints, and skills associated with the effective monitoring of resources can be crippling to management efforts (Savage et al., 2017). Orthodox methods of monitoring both biodiversity and resource use are often costly and hard to sustain, especially where financial resources are limited (Danielsen et al., 2005). This adds to the factors contributing to the implementation gaps listed above, contributing to the risk of paper-park establishment, further highlighting that we need a better way.

Making environmental governance good

Given that our environment is a self-governing system, it would be easy to assume that no governance would be the best governance. However, if we revisit our earlier discussion about the Earth's carrying capacity for humans being artificially increased through the implementation of technology, and the degree of environmental decline that has already been identified (Myers et al., 2025) it becomes apparent that we are past the point of expecting the environment to self-govern. Environmental policies are under a tremendous amount of pressure to deliver success. However, we need to reframe the discussion around environmental policy to acknowledge that environmental policy is not about the environment, but is instead about managing the way that humans interact with the environment. The difference here is subtle but important, because it re-centres humans and human behaviour as the focus of the policy, rather than on the periphery.

This is critical because our current approach serves neither our environment nor the humans that are depending on the resources it produces. There are often cases where the implementation of management strategies can feel imbalanced in terms of ecological versus social need:

> You cannot tell a fisher in a developing nation that they must throw a fish back into the ocean because it is too small, or that they must only catch four fish per day when they have eight family members to feed.
>
> (Russ, 2002, 421)

Fishing quotas and size limits are an excellent example of a situation where environmental policy does not necessarily take into account localised socio-economic needs and dynamics. The implementation of a minimum catch size is based on the scientific principle that larger fish are generally older, and therefore more likely to have met reproductive age, adding to the population, before they themselves are subtracted. As such, the development of minimum size limits as a policy instrument has a clear environmental connection, but fails to recognise human socio-economic needs, which is ironic, given that the purpose of such a policy is to manage the human extraction of fish. Fish cannot apply these regulations to themselves.

Local contexts can be as diverse as the environments they are designed to protect, and unfortunately, the top-down/trickle-down mechanisms for local-level policy implementation can sometimes be too varied for individual countries to manage. This results in more of a one-size-fits-all approach (Agardy et al., 2003) and often fails to facilitate the incorporation of social impacts across different social scales. All of which suggests the need for a more interdisciplinary approach in the design of policy targets.

Supranational policies are intended to work in a top-down manner, triggering subsequent policy development at lower levels of the policy hierarchies at regional, national, and/or local levels. In some ways this is sensible, as it facilitates adaptable implementation which is, at least in theory, intended to be flexible to the background and needs of specific local contexts. However, the reality in which a policy manifests can be quite different. One of the key factors limiting the success of any management strategy is the availability of resources. There is a general consensus that we lack both the financial resources and manpower to meet the demand set by supranational policy targets (Uychiaoco et al., 2005; Danielsen et al., 2014; Savage et al., 2017, 2020). Clearly, there is a need for more accessible monitoring and management methods (Savage et al., 2020) and preferably ones that more comprehensively integrate local communities and their knowledge. In recent years there has been a progressive shift towards the integration of local community members and volunteers to help plug knowledge and skills gaps, but there is a lot of scepticism in scientific circles about the reliability and quality of data collected by 'non-professionals' (Danielsen et al., 2014; Savage et al., 2017). This ties in to the discussion that follows in Chapter 5, where it is argued that European-style governance systems whitewash alternative worldviews about human-nature relationships under the guise of 'lack of capacity.' Overall, implementation issues suggest that our targets are too big to be achievable, and perhaps it is time to scale back our expectations, and focus on quality over quantity.

Breaking out of silos and moving forward

During his tenure as Secretary General of the United Nations, Ban Ki Moon said 'we can no longer afford to think and work in silos' during one of his speeches launching the SDGs in 2015. This statement had the aim of highlighting the intended interdisciplinarity of the SDGs, a great step forward in many ways. However, while awareness of the need for interdisciplinarity in targets is increasing, the breadth and complexity of the targets are also increasing.

Supranational policy commitments intended to support environmental protection are challenging for governments to achieve. Targets may be met on paper, for example, in terms of area protected, and relatively 'easy' to implement politically, but could be misleading in terms of efficacy (Relano & Pauly, 2023). In some cases, targets have been useful, but often interventions made to fulfil targets not only falsely suggest progress or success, but they may actively impede future conservation efforts (Agardy et al., 2016). Effective implementation of protected areas requires significant buy-in and compliance from local resource users. In order to encourage both, managers will often suggest that protected area benefits will be spread across all stakeholders, but environmental recovery takes time. For example, it is typically assumed that fisheries require effective protection for 15–40 years before fish populations could be expected to recover (Savage et al., 2020). This is a long time for people to wait, and appropriate provision needs to be in place to support people in the meantime.

If we look at examples of success within the environmental policy arena, our early attempts appear to have been more successful. Supranational policies such as the Ramsar Convention, which successfully reduced rates of environmental degradation in wetlands, show that implementation can be effective. The 1987 Montreal Protocol, which was able to restore ozone integrity, is another example in this regard. However, these conventions were more specific, targeted on specific issues or habitats, and with clear indicators of what success would look like. On the other hand, as discussed throughout this chapter, policies which are too big, or too complex in terms of targets and implementation mechanisms tend to struggle.

Evidence of environmental degradation is everywhere at the moment, and it makes sense that from an 'optics' perspective we need to be demonstrating that our collective efforts are working to respond to the environmental polycrisis. Understandably, no one wants to admit that current approaches are not suitable, because no one wants to 'own' that failure, and while a degree of flexibility in the design and implementation of governance structures is essential, too much flexibility risks diluting effort to the point of failure. Complex supranational policy instruments are not working. Initiatives such as the MDGs, SDGs, CBD, and Paris Agreement are too complex, and offer too many routes towards sustainable development for individual nations, to the point that they limit global movement towards success.

So where do we go from here? We learn from our past successes, and go back to basics. We must re-centre environmental policy to explicitly address the human component, focusing governance tools on human-ecosystem interactions and human use of the environment, rather than on the environment itself. We need to fully integrate interdisciplinary thinking into the development of smaller, tangible, and achievable targets, making sure that we meet the aims of one target before prioritising the next, in a step-by-step approach. Perhaps, however, the greatest step towards good environmental governance would be the ability and humility to acknowledge when things are not working, to take a step back, rethink, and then move forward.

References

Agardy, T., Bridgewater, P., Crosby, M. P., Day, J., Dayton, P. K., Kenchington, R., Laffoley, D., McConney, P., Murray, P. A., Parks, J. E. & Peau, L. 2003. Dangerous targets?: Unresolved issues and ideological clashes around marine protected areas. *Aquatic Conservation: Marine and Freshwater Ecosystems* 13: 353–367.

Agardy, T., Claudet, J. & Day, J.C. 2016. 'Dangerous targets' revisited: Old dangers in new contexts plague marine protected areas. *Aquatic Conservation: Marine and Freshwater Ecosystems* 26(2): 7–23. https://doi.org/10.1002/aqc.2675

Barichievey, C., Munro, L., Clinning, G., Whittington-Jones, B. & Masterson, G. 2017. Do armed field-rangers deter rhino poachers? An empirical analysis. *Biological Conservation* 209: 554–560.

Blais, F. 2002. The fair and equitable sharing of benefits from the exploitation of genetic resources: A difficult transition from principles to reality. In Le Prestre, P. G. (Ed.). *Governing Global Biodiversity: The Evolution and Implementation of the Convention on Biological Diversity* (1st ed., pp. 145–158). Routledge, London. https://doi.org/10.4324/9781315253930

Brundtland, G. 1987. Report of the World Commission on Environment and Development: Our Common Future. United Nations General Assembly document A/42/427.

Cohen, J. E. 1995. Population growth and earth's human carrying capacity. *Science* 269(5222): 341–346.

Cox, M., Villamayor-Tomas, S., Epstein, G., Evans, L., Ban, N.C., Fleischman, F., Nenadovic, M. & Garcia-Lopez, G. 2016. Synthesizing theories of natural resource management and governance. *Global Environmental Change* 39: 45–56.

Cumming, G. S. 2025. Conceptualizing and measuring ecological spillover effects from protected areas. *Frontiers in Ecology and Evolution* 2025: e70008. https://doi.org/10.1002/fee.70008

Danielsen, F., Burgess, N. D., & Balmford, A. 2005. Monitoring matters: Examining the potential of locally-based approaches. *Biodiversity Conservation* 14: 2507–2542.

Danielsen, F., Jensen, P. M., Burgess, N. D., Altamirano, R., Alviola, P. A., Andrianandrasana, H., Brashares, J. S., Burton, A. C., Coronado, I., Corpuz, N., Enghoff, M., Fields, J., Funder, M., Holt, S., HUbert, H., Jensen, A. E., Lewis, R., Massao, J., Mendoza, M. M., Naga, Y., Piper, C. B., Poulsen, M. K., Rueda, R. M., Sam, M. K., Skielboe, T., Sorensen, M. & Young, R. 2014. A multicountry assessment of tropical resource monitoring by local communities. *Bioscience* 64(3): 236–251.

Fehling, M., Nelson, B. D. & Venkatapuram, S. 2013. Limitations of the Millennium development goals: A literature review. *Global Public Health* 8(10): 1109–1122. https://doi.org/10.1080/17441692.2013.845676

Francioni, F. 2014. Foreword. In Morgera, E., Tsiuomani, E. & Buck, M. (Eds.). *Unraveling the Nagoya Protocol: A Commentary on the Nagoya Protocol on Access and Benefit-Sharing to the Convention on Biological Diversity* (pp. viii–xvi). Brill, Leiden.

Gardner, R. C. & Davidson, N. C. 2011. The Ramsar Convention. In LePage, B. (Ed.). *Wetlands*. Springer, Dordrecht. https://link.springer.com/chapter/10.1007/978-94-007-0551-7_11

Graham, J., Amos, B. & Plumptre, T. 2003. *Governance Principles for Protected Areas in the 21st Century.* Institute on Governance, Parks Canada and the Canadian International Development Agency, Ottawa.

Howlett, M. & Rayner, J. 2006. Convergence and divergence in 'new governance' arrangements: Evidence from European integrated natural resource strategies. *Journal of Public Policy* 26: 167–189.

Jones, J. P. G., Collen, B., Atkinson, G., Baxter, P. W. J., Bubb, P., Illian, J. B., Katzner, T. E., Keane, A., Loh, J., McDonald-Sommerville, M. & Milner-Gulland, E. J. 2011. The why, what and how of global biodiversity indicators beyond the 2010 target. *Conservation Biology* 25(3): 450–457 https://doi.org/10.1111/j.1523-1739.2010.01605.x

Lemos, M. C. & Agrawal, A. 2006. Environmental governance. *Annual Review of Environment Resources* 31: 185–194.

Le Prestre, P. G. (Ed.). 2002. *Governing Global Biodiversity: The Evolution and Implementation of the Convention on Biological Diversity* (1st ed.). London: Routledge. https://doi.org/10.4324/9781315253930

Lockwood, M., Davidson, J., Curtis, A., Stratford, E. & Griffith, R. 2010. Governance principles for natural resource management. *Society and Natural Resources* 23(10): 986–1001.

Ma, F., Wang, H., Tzachor, A., Hidalgo, C. A., Schandl, H., Zhang, Y., Zhang, J., Chen, W-Q., Zhao, Y. Shu, Y-G. & Fu, B. 2025. The disparities and development trajectories of nations in achieving the sustainable development goals. *Nature Communications* 16: 1107. https://doi.org/10.1038/s41467-025-56076-6

McGraw, D. 2002. The story of the biodiversity convention: From negotiation to implementation. In Le Prestre, P. G. (Ed.). *Governing Global Biodiversity: The Evolution and Implementation of the Convention on Biological Diversity* (1st ed.). Routledge, London. https://doi.org/10.4324/9781315253930

Morgera, E., Tsiuiomani, E. & Buck, M. 2014. *Unraveling the Nagoya Protocol: A Commentary on the Nagoya Protocol on Access and Benefit-Sharing to the Convention on Biological Diversity.* Brill, Leiden.

Myers, S. S., Masztalerz, O., Ahdoot, S., Gabrysch, S., Gupta, J., Haines, A., Kleineberg-Massuthe, H., Lambrecht, N. J., Landrigan, P. J., Mahmood, J., Pörtner, L. M., Rohr, J., Traidl-Hoffmann, C., Wendt A. S., Wray, B. & Rockström, J. 2025. Connecting planetary boundaries and planetary health: A resilient and stable Earth system is crucial for human health. *The Lancet* 406(10501): 315–319.

Natsir, M., Ruchimat, T., Agustina, S. & Yulianto, I. 2019. Application of global positioning system tracker to detect the fishing ground location and effective effort in Artisanal Fishery. *Sensors and Materials* 31(3): 803–814.

Pierre, J. 2000. Introduction: Understanding governance. In Pierre, J. (Ed.). *Debating Governance* (pp. 1–10). Oxford University Press, New York.

Pythoud, F. & Thomas, U. P. 2002. The Cartagena protocol on biosafety 1. In Le Prestre, P. G. (Ed.). *Governing Global Biodiversity: The Evolution and Implementation of the Convention on Biological Diversity* (1st ed., pp. 39–56). Routledge, London. https://doi.org/10.4324/9781315253930

Rattan, R., Sharma, B., Kumar, R., Saigal, V. & Shukla, S. 2021. Ramsar convention. In Sharma, S. & Singh, P. (Eds.). *Wetlands Conservation* (pp. 17–39). Wiley, Hoboken. https://doi.org/10.1002/9781119692621.ch2

Relano, V. & Pauly, D. 2023. The 'Paper Park Index': Evaluating Marine protected area effectiveness through a global study of stakeholder perceptions. *Marine Policy* 151: 105571. https://doi.org/10.1016/j.marpol.2023.105571

Roberston, P. J. & Choi, T. 2010. Ecological governance: Organizing principles for an emerging era. *Public Administration Review* 70(1): s88–s99. https://doi.org/10.1111/j.1540-6210.2010.02250.x

Rockström, J., Steffen, W., Noone, K. et al. 2009. A safe operating space for humanity. *Nature* 461: 472–475. https://doi.org/10.1038/461472a

Rodriguez-Rodriguez, D., Bomhard, B., Butchart, S. H. M. & Foster, M. N. 2011. Progress towards international targets for protected area coverage in Mountains: A multi-scale assessment. *Biological Conservation* 114(2011): 2978–2983.

Rosenau, J. N. 2000. Change, complexity, and governance in globalizing space. In Pierre, J. (Ed.). *Debating Governance* (pp. 167–200). Oxford University Press, New York.

Russ, G. R. 2002. Yet another review of marine reserves as reef fish management tools. *Coral Reef Fishes*: 421–443.

Sachs, J. 2012. From Millenium development goals to sustainable development goals. *The Lancet* 379: 2206–2011.

Savage, J. M., Hudson, M. D. & Osborne, P. E. 2020. The challenges of establishing marine protected areas in South East Asia. In Humphreys, J. & Clark, R. W. E. (Eds.). *Marine Protected Areas* (pp. 343–359). Elsevier, Amsterdam. https://www.sciencedirect.com/science/chapter/edited-volume/abs/pii/B9780081026984000186

Savage, J. M., Osborne, P. E. & Hudson, M. D. 2017. Can international volunteers and community members effectively monitor coral reef resources in Cambodia? *Aquatic Conservation: Marine and Freshwater Ecosystems* 27(2): 340–352.

Schugart-Schmidt, K. L., Pike, E. P., Moffitt, R. A., Saccomanno, V. R., Magier, S. A. & Morgan, L. E. 2015. SeaStates G20 2014: How much of the seas are G20 nations really protecting? *Ocean and Coastal Management* 115: 25–30.

Siegel, F. R. 2021. *The Earth's Human Carrying Capacity: Limitations Assessed, Solutions Proposed*. Springer Nature, Cham.

Singer, F. D. 2016. *Ecology in Action*. Cambridge University Press, Cambridge.

Toropova, C., Meliane, I., Laffoley, D., Matthews, M. E., & Spalding, M. 2010. *Global Ocean Protection: Present Status and Future Possibilities*. IUCN, Gland, Switzerland.

United Nations 2015. *Transforming our world: The 2030 Agenda for Sustainable Development*. https://sdgs.un.org/2030agenda

Uychiaoco, A. J., Areceo, H. O., Green, S. J., Margarita, T., Gaite, P. A., & Aliño, P. M. 2005. Monitoring and evaluation of reef protected areas by local Fishers in the Philippines: Tightening the adaptive management cycle. *Biodiversity and Conservation* 14(11): 2775–2794.

Van Leeuwenhoek, A. 1679. Letter 43: A van Leeuwenhoek to N. Grew. In A Committee of Dutch Scientists (Eds.). *Alle de brieven van Antoni van Leeuwenhoek/The Collected Letters of Antoni van Leeuwenhoek* (Vol. 3, pp. 4–35). Swets & Zeitlinger, Amsterdam.

Visconti, P., Butchart, S. H. M., Brooks, T. M., Langhammer, P. F., Marnewick, D., Vergara, S., Yanosky, A. & Watson, J. E. M. 2019. Protected area targets post-2020. *Science* 364(6437): 239–241.

White, T. D., Ong, T., Ferretti, F., Block, B. A., McCauley, D. J., Micheli, F. & De Leo, G. A. 2020. Tracking the response of industrial fishing fleets to large marine protected areas in the Pacific Ocean. *Conservation Biology* 34(6): 1571–1578.

Zelnik, Y. R., Arnoldi, J- R. & Loreau, M. 2018. The impact of spatial and temporal dimensions of disturbances on ecosystem stability. *Frontiers in Ecology and Evolution* 6: 224. https://doi.org/10.3389/fevo.2018.00224

3

GOOD LABOUR GOVERNANCE

Maria Gavris and Jason Heyes

Introduction

Labour governance, or the governance of work, is central to sustainable development. This is because work is the mediating link between society and the environment (Kreinin and Aigne, 2022). For humanity to stay within planetary boundaries while also meeting the necessary social foundations for development, both qualitative and quantitative changes in the world of work are required. We will need to reduce work in certain sectors (the polluting and extractive industries), increase it in others (the care sector and public services), and perhaps create new forms of work altogether, all of which will require changes in how labour is governed.

Good labour governance is therefore essential for sustainability. However, it remains an under-researched topic, much like good governance itself, with ambiguous and contested definitions. Labour governance is sometimes equated with the regulation of formal employment, a view that reflects the realities of affluent Global North countries but does not adequately capture the diversity of work arrangements in the Global South (Meardi and Marginson, 2014). More recent definitions adopt a broader and more inclusive perspective. LeBaron (2020, 3), for instance, defines labour governance as 'the public and private standards regulations, responses and forms of power (including rules, norms and actions) that surround labour standards in the global economy.' This definition is intended to cover transnational private regulation by non-state actors, including corporate social responsibility, 'state-based regulation, enforcement and power relations,' and international labour standards. These are undoubtedly important components of labour governance. However, also of importance to labour governance are the processes through

DOI: 10.4324/9781003467694-4

which labour policies are created and implemented. These aspects of labour governance relate to what the International Labour Organisation (ILO) terms 'labour administration.' Important considerations in this regard are the ability of governments to coordinate and support policy development and the extent to which non-state actors are consulted about policy developments or reforms and have opportunities to influence their content. National governments play a fundamentally important role in labour governance through, for example, creating the framework of labour laws, regulations, social protections, and institutions (e.g. social dialogue institutions) within which work and employment actors operate and interact and providing the means by which rights might be enforced (particularly by establishing and funding labour inspection services and labour courts). Governments shape the scope for other actors – primarily trade unions, and employer organisations, as well as, potentially, other civil society organisations (CSOs) – to participate in labour (and labour market) governance. These forms of participation can include engaging in tri- and bi-partite social dialogue (Papadakis, 2021), involvement in actions to strengthen the enforcement of labour standards and, in some countries, playing a role in the administration of bodies responsible for matters such as pensions and vocational training.

Within federal systems matters are more complex as authorities within provinces/states may have the autonomy to adopt their own regulations, raising questions relating to subsidiarity and whether effective 'multilevel' governance can be ensured. Considerations relating to multilevel governance also extend to the relationship between national and international dimensions of governance. This is evident in the European Union (EU) where European integration has led to the creation of new vertical relationships that involve the European Commission and EU member states, and new horizontal relationships that involve coordinated actions by employment relations actors in different member states (Marginson and Sisson, 2004). More widely, the ILO is obviously a major actor in global labour governance through its role in establishing international labour standards, monitoring their observance by member countries and providing technical assistance to employer organisations, trade unions, and national governments.

What does good labour governance entail? Existing literature offers little explicit guidance on this question, which is perhaps unsurprising given that 'labour governance' itself is a contested and complex concept – even before we add the normative weight of the term *good*. Nevertheless, the question is crucial if we aim to build a sustainable world. In some policy circles (e.g. the World Bank), good labour governance is treated as indistinguishable from good governance more broadly, with the assumption that the same principles apply across domains – whether in labour, trade, finance, etc. However, this perspective falls short in capturing the specific role that labour governance can and should play in sustainable development. It is also essential to recognise the interconnection between labour governance and environmental issues.

Implicitly, for the ILO key ingredients of good labour governance would seem to include clear labour laws that underpin the promotion of decent work, effective labour inspection and public employment services, reliable dispute-settlement mechanisms and participation by employer organisations and trade unions in governance. The first part of this chapter examines some of the challenges that countries face in this regard, which range from the exclusion from labour governance of various 'non-standard' forms of work to challenges in the enforcement of labour standards, especially on a global level. We then place these issues within the broader context of the climate crisis, exploring relevant theories that can shed light on the relationship between labour governance and sustainability. We distinguish between the green growth paradigm and the alternative degrowth approach. Engaging with these theories allows us to explore the role labour might play in a sustainable world and the implications of this for labour governance, while also highlighting some gaps in the existing literature. In the conclusion, we build on this analysis to articulate an emerging answer to the question 'What makes labour governance good?'

The current state of labour governance: challenges

The coverage of labour laws and social protection

Possibly the most significant challenge facing labour governance relates to its coverage. Around 2 billion of the world's employed population (almost 60% of all workers including those in agriculture and 50% excluding agriculture) work in the informal economy (ILO, 2023: 15). These workers are typically excluded from labour law and social protections and experience vulnerability, low pay and poor working conditions (OECD/ILO, 2019; Rogan et al., 2018). Informal jobs are most common among own-account workers: 87% of own-account workers and 55% of employers own and operate an informal economic unit. Workers in forms of non-standard employment other than self-employment, such as fixed-term or part-time employment, are also more likely to be in the informal economy when compared with workers in 'standard' employment (ILO, 2023). Informal employment is also particularly common in some occupations. For example, the ILO (2021) has estimated that more than 81% of domestic workers are informally employed. Although the ILO has emphasised that domestic workers should be properly recognised as workers, and ILO Convention 189 establishes that domestic workers caring for families and households should have the same basic labour rights as other workers, most are effectively excluded from labour governance. Even when they are granted rights and protections, the fact that domestic workers carry out their jobs within private households makes the task of enforcing standards particularly problematic.

Self-employed workers in the formal economy are also typically excluded from the coverage of labour laws and social protections on the basis that their work is not performed in the context of an employment relationship. In practice, however, it is not always easy to determine which workers are genuinely self-employed. While many legal systems involve a 'binary divide' between employment and independent self-employment (Countouris and Freedland, 2013), between these two conditions is a grey area that includes bogus self-employment (or disguised employment) and dependent self-employment. The former situation is one in which 'the worker is purposefully misclassified as independent, self-employed worker, even though he or she is, in fact, in a subordinate employment relationship' (ILO, 2016, 9). In the latter situation, 'workers perform services for a business under a civil or commercial contract but depend on one or a few clients for their income or receive direct instructions on how the work is to be carried out' (ILO, 2016, 9). Other types of employment can also be excluded or insufficiently covered by labour laws or social protections. For example, the social protections extended to casual workers are often very limited and workers in part-time employment can be excluded from public social insurance schemes because they do not meet legal thresholds relating to minimum working time, employment duration, or earnings (Eurofound, 2019).

Participation in labour governance

Social dialogue can make an important contribution to labour governance. The term refers to the 'the involvement of workers, employers and governments in decision-making on employment and workplace issues' (Papadakis, 2021, 43). It can be either bi-partite, involving only employers and workers, or tripartite if it also includes governments. Beyond tripartite social dialogue, there might be opportunities for civil society groups other than employers' and workers' organisations to be consulted and contribute to activities related to labour administration (referred to by the ILO as 'tripartite-plus' processes). CSOs have, for example, contributed to the design and implementation of standards and various reporting, auditing, monitoring, verification, and certification systems (ILO, 2024, 125).

Most ILO member countries (around 87%) have established permanent institutions to enable general social dialogue and around half have bodies relating to specific issues such as employment (ILO, 2024). Social dialogue has proved extremely important in times of economic and social crisis, such as the 2008 financial crisis and its aftermath and the COVID-19 pandemic, when social dialogue in some countries facilitated the introduction of measures to support jobs and businesses (Brandl, 2025). At the same time, crises can place social dialogue under strain. For example, in the European countries hardest hit by the financial crisis, such as Spain and Ireland, external conditions attached to bailout packages reduced domestic room for manoeuvre as well as timeframes for consultation with social partners (Hyman, 2013; Gavris and Heyes, 2019).

Social dialogue faces serious challenges in many parts of the world. These include falling membership among employers' organisations and trade unions, a growth in new forms of work involving workers who often lack collective representation, growing inequality, and a 'decline in the use of forms of social dialogue that produce binding commitments' (Papadakis, 2021, 45). Women, young people, and those who work in the informal economy tend to be under-represented in social dialogue fora (ILO, 2024). In some countries, basic necessary preconditions for social dialogue, such as political stability and freedom of association, continue to be absent. Even where mechanisms for consultation, cooperation, and negotiation exist, they are sometimes non-operational, ineffective, or in need of reform (ILO, 2024, 119). This can be, for example, because meetings are not held, consultation takes place very infrequently, insufficient financing is provided to enable social dialogue, or the views of social partners are disregarded by ministries when formulating or revising policies.

Fissuring

Business strategies that involve the reorganisation and decentralisation of production and service delivery present significant labour governance challenges. Common examples of these strategies, which Weil (2014) refers to as 'fissuring,' are subcontracting and outsourcing. Both involve allocating jobs outside of a firm. Outsourcing involves allocating entire jobs or departments to an external firm, often on a permanent basis. Subcontracting involves hiring an external firm or individual to undertake a task on a short- or long-term basis. An example of the effects of subcontracting is provided by Lucassen (2021: 431–432):

> In 1974 the Dutch electronics company Philips had almost half a million employees compared to 37,000 now. [...] The explanation for this kind of development is deeply rooted in the proliferation of subcontracting. And where many wage workers are still physically working together, for example in the gigantic distribution centres located alongside motorways which are rapidly replacing shops in town centres, this is happening not only as a result of endless subcontracting but also as a result of payroll constructions.

Outsourcing and subcontracting can enable employers to lower their costs, evade the provisions of collective agreements, and transfer the risks associated with fluctuating workloads to external staff (Jaehrling and Méhaut, 2013; Weil, 2014). Activities may be further fragmented via additional subcontracting so that lead employers are 'orbited' by successive tiers of business enterprises, with relationships between the tiers being specified in separate contracts or agreements (Weil, 2014). The fragmentation of work can result in a lack of transparency regarding the identity of the employer (Durán and Narbona, 2021). In other words, workers can be uncertain who they 'really work for.'

The use of third-party labour intermediaries or labour contractors/brokers creates a further type of multiple-party employment relationship, wherein the legal employer is separate from the organisation for which the work is performed. Companies may hire formerly directly employed workers through employment agencies, providing worse conditions of employment (Borges Neves and Krekula, 2023). In some countries, such as India, intermediaries have played a significant role in facilitating a growth of informal employment in enterprises in the formal sector (Srivastava, 2016). Complex subcontracting chains can encourage the growth of bogus self-employment, which is prevalent in labour intensive industries such as transport, construction, business services, and cleaning (Thörnquist, 2013; Wagner and Bernsten, 2016). Labour intermediaries and elaborate subcontracting chains are common in these industries and contribute to bogus self-employment by hiding workers' true employment status (Thörnquist, 2013).

Enforcement

Labour governance remains primarily the domain of national-level institutions. However, national labour administration systems – comprising institutions such as labour ministries, labour inspectorates, and public employment services – are often underfunded, understaffed, and weakened by competing priorities from other national institutions, such as finance ministries (Gavris and Heyes, 2019). This makes enforcement of labour standards difficult. The picture becomes even more complex if we consider transnational labour issues, which cannot be adequately addressed at the national level. At the global level, we have a global goal, SDG8 Economic Growth & Decent Work, but no global institutions with the power to enforce it. A separate issue is that this goal subsumes 'decent work' under economic growth, signalling that work is prioritised in global governance only insofar as it contributes to economic growth.

At the national level

The enforcement of employment protections has become more problematic in many countries. One important factor is the shrinking coverage of trade unions. Krings' (2014) analysis of construction and hospitality sectors in Ireland shows that employers have frequently engaged in 'social dumping' (e.g. sought to undermine or evade regulations), in different ways depending on the specific regulatory environment and whether trade unions are present. In hospitality, where unions and collective bargaining have been largely absent, low-wage jobs and casualisation have proliferated, and social dumping has involved non-compliance with employment regulations. Trade unions have, by contrast, been relatively strong in the construction sector, with industry-wide agreements regulating pay and conditions. Nevertheless, employers have still been able to engage in social dumping through increased use of subcontracting arrangements, with resulting employment fragmentation and informalisation down the subcontracting chain.

In many countries, the ability of labour inspectorates to enforce labour standards remains weak. The resources available to inspectorates are often insufficient for them to carry out their functions effectively, a problem that is prevalent and severe in the Global South (ILO, 2011; Casale and Fasani, 2012). Many inspectorates struggle with issues such as insufficient staff and material resources, inadequate training of staff at all levels, low staff motivation due to inadequate compensation and career progression, and lack of support from other parts of government for compliance with labour laws and mechanisms to coordinate compliance efforts. Moreover, in some countries, fiscal retrenchment in the wake of the 2008 financial crisis led to reductions in the budgets and hence the capacity of labour inspectorates (Gavris and Heyes, 2019).

On a global scale

The creation of complex cross-border supply chains has added to governance challenges (Fransen and Burgoon, 2015; Marx and Wouters, 2016). In discussing potential sources of pressure for social upgrading within value chains, Gereffi and Lee (2016) stress the importance of interactions between place-based, horizontal forms of governance that operate within industrial clusters and cross-national, vertical governance which operates along the length of value chains. In both cases, there are elements of private governance, which involves purchasers, suppliers and other private sector actors, social governance, which involves CSOs and unions, and public governance, which includes state actors at different levels. Research has pointed to the inadequacy of multinational corporations' (MNCs) private governance arrangements involving codes of conduct and supplier audits (for a review of evidence, see Li and Hu, 2022). Problems include the fact that audits are typically conducted by a small number of auditors (often a single auditor) at a given point in time, as opposed to being a continuous process of monitoring. Moreover, auditing approaches face substantial challenges. Not only do multinationals often have many suppliers, but those suppliers then further outsource work to, for example, homeworkers, as often occurs in the garment and textile industry (Marx and Wouters, 2016). In addition, factories that are facing pressure to comply with labour standards might close and relocate elsewhere. An additional limitation is that audits tend to be better at assessing standards that can be relatively easily measured and evidenced, such as wages and health and safety, than more complex issues such as freedom of association or gender discrimination (Barrientos et al., 2019; Barrientos and Smith, 2007).

These difficulties have led to calls for continuous monitoring and 'bottom up' initiatives that involve local stakeholders (Marx and Wouters, 2016). Some studies (Coslovsky, 2014; Locke, 2013; Locke et al., 2013) have found that private and public actors can combine to promote compliance, although MNCs' compliance initiatives rarely set out to involve labour inspectorates (Barrientos and Smith, 2007; Locke et al., 2009; Rossi, 2015). Cooperation between CSOs and trade unions can also help to improve standards in global supply chains, as appears to have been the case in the Cambodian garment export industry (Li and Hu, 2022) and Kenya's tea and flower industries, where the involvement of CSOs in 'social governance' has led to initiatives to tackle issues such as sexual harassment (Barrientos et al., 2019). The ILO's Better Work Programme, launched in 2007 in cooperation with local stakeholders and covering more than 1 million workers in the garment industry in the Global South is another important example of a successful programme improving compliance (Rossi, 2015).

Labour governance in the context of the climate crisis: green growth versus degrowth

Alongside the issues raised above, another problem with labour governance until recently has been the limited regard for how it interacts with environmental aspects of sustainability. However, recurring intersecting crises have led policymakers and scholars to recognise some of the negative implications of economic growth and devise alternative development models to address the environmental and resource challenges of the 21st century. In this context, green growth emerged as a popular governance paradigm, which promises to reconcile economic growth and environmental sustainability.

The key aspects of green growth are a shift away from oil dependency and a switch to renewable energy sources; the emphasis on new, cleaner technologies to improve energy efficiency; a move away from manufacturing to services; waste management; and the creation of more secure well-paying jobs. This latter point in particular appeals to the labour movement. For example, the 2018/2019 proposal for a Green New Deal in the US included a Job Guarantee to give 'every unemployed American who wants one, a job building energy-efficient infrastructure' (Pettifor, 2019). Green growth aided by technology is thus being presented as a 'win-win-win solution for employment, environment, and development' (Saave and Muraca, 2021, 745).

Scholars in the degrowth movement contest this, however. For instance, Kallis (2015) claims that the emphasis on greater efficiency creates a paradox, because under a capitalist consumerist logic, freeing up resources in one sector of the economy through energy-saving technologies would just lead to more consumption elsewhere in the economy. Also, the switch to services has not led to a decline in material footprint, perhaps because some services (for instance retail, and especially deliveries) are still quite resource-intensive and polluting. Social justice between the Global North and the Global South is another important issue raised by degrowth scholars. For example, the switch to electric vehicles is considered environmentally sustainable in the Global North, but it in fact relies on critical raw materials extracted from 'sacrifice zones' in the Global South, with significant negative impacts on the environment. For these reasons, degrowth scholars stress that there is no evidence to support the possibility of decoupling GDP growth from resource use and carbon emissions on a global scale (Hickel and Kallis, 2020).

As a result, there is growing recognition in the sustainability literature that the size of the economy will need to adjust to ecological limits. In practical terms, this will mean some form of degrowth (Zekić, 2024). With the caveat that degrowth literature is complex and consists of different strands, a succinct definition of degrowth is a planned and voluntary downscaling of the current size and shape of our economies, to keep them in balance with the natural world. Degrowth is not about finding a 'greener' way to produce, but about recognising the need to downscale production altogether and focus on maintaining what we have instead. It is a post-capitalist perspective which emphasises sufficiency over efficiency.

While the implications of green growth for labour are clearly spelled out (more jobs in the green sectors), degrowth is more ambiguous on this point. Nevertheless, it is possible to excavate some core points about the role of labour in a sustainable world from the key degrowth debates.

What role for labour in a sustainable world?

Two main changes in the world of work are identified as essential to a sustainable transition in degrowth theory (Vincent and Brandellero, 2023). The first is a change in working time to match a shift away from the profit-oriented market economy and a reduction in wasteful production and consumption. The second is a need to rethink what we mean by 'work,' what types of work matter, and what work is actually for. Both are closely connected as part of the project to re-embed the economy in society.

Changes in working time

At the heart of degrowth lies the proposition that we need to cut wasteful expenditure and eliminate or shrink polluting, extractive, or otherwise socially unnecessary (e.g. advertising) sectors. Without compensatory policies in place, such a change would lead to mass unemployment. To avoid this, degrowth scholars advocate for a redistribution of work in society through a shorter work week, coupled with a universal basic income and job guarantees:

> As we shed unnecessary jobs we can shorten the working week, going from forty-seven hours (the average in the United States) down to thirty or perhaps even twenty hours, distributing necessary labour more evenly among the working population and maintaining full employment. We can facilitate this process by introducing a job guarantee (…), and roll out retraining programmes so that people laid off from shrinking industries can transition easily to others.
>
> (Hickel, 2020, 83)

A reduction in working time, coupled with a redistribution of work in society and a universal basic income is expected to bring triple benefits: lowering unemployment, reducing pressure on the environment, and liberating workers from oppression and allowing them to pursue other activities not oriented towards production, which improve their quality of life, such as spending time with loved ones or caring for the environment.

Reimagining the role of work in society

Discussions around shedding unnecessary jobs imply a pressing need to define what we mean by 'necessary' and 'unnecessary' labour, and what labour is for. Quantitative changes are thus inextricably linked to qualitative changes in the world of work.

On the one hand, degrowth principles implicitly urge us to identify and eliminate what David Graeber (2018) calls 'bullshit jobs': 'a form of paid employment that is so completely pointless, unnecessary, or pernicious that even the employee cannot justify its existence' (2018, 9–10). These are jobs that, were they to disappear tomorrow, would make no substantive difference in society and might even make the world a better place. According to Graeber, jobs such as 'lobbyist,' 'corporate lawyer,' and 'middle manager' use resources without generating social value and were designed to keep people occupied and therefore distracted from recognising the problems of the capitalist system and revolting against it.

On the other hand, there are types of work whose existence is vital to nature and society, but which are often rendered invisible and devalued in the growth economy. Meta-industrial labour, a term coined by eco-feminist scholar Ariel Salleh (1997), refers to unpaid work which mediates human and natural cycles to maintain the necessary biological infrastructure for all economic systems ('reproductive work'), performed by women in households, by small-scale farmers in rural areas, and by indigenous groups. This work is highly gendered, racialised, and unrewarded; as noted above these are precisely the categories of workers who are excluded from legal coverage, social protections, and participation in labour governance through social dialogue. Degrowth scholars argue that a sustainable society must place care at its core. Hickel (2020), for example, argues that the most important part about a shorter work week is that it frees people to spend more time on what really matters and directly contributes to social and ecological well-being and therefore happiness: care – for the infirm or elderly, for children, for nature, for each other.

Implications for labour governance

There is a tension in the degrowth literature between what appears to be an anarchist strand and a 'bottom-up governance with state support' strand. Some believe that a degrowth vision for labour can only be implemented on a local level, because the state is too entangled in the growth mindset and with the capitalist system. In this view, grassroots-based direct democracy must subvert the authority of the state to create 'spaces in which we can lead our lives autonomously, with different rules of the game than those imposed on us by the system' (Pérez Orozco, 2022, 233). Questions around what labour is for and who should perform it would happen at the level of small local communities. Waged labour would be replaced with democratically defined and shared 'ecologically appropriate work within local geographies of community sufficiency' (Nelson, 2024, 61).

While the potentials of local governance should not be underestimated, others have pointed out the limitations of this approach. One criticism is that if the state serves the interests of capital, then this raises the question of what mechanisms exist to prevent the state from simply crushing these emerging communities seeking an exit from capitalism. Another point of criticism is that:

> One of the strategic blind spots of such thinking is that human societies must deal with the legacies of the past. The current generation must deal with the existing production/consumption nexus. It cannot get rid of this overarching socialization (...) by simply retracting from intensely connected metropolis to form small local communities. The transition beyond growth also needs to be addressed materially and institutionally (e.g. by phasing out fossil-based or otherwise dirty or unnecessary activities) at a level inaccessible to local actors.
>
> (Durand et al., 2024, 3)

In other words, the degrowth transition cannot start from a clean slate, but must address existing power relations and structural barriers to sustainability. Indeed, while arguing that a degrowth transition will be heavily bottom-up, Kallis (2015) nonetheless contends that this cannot succeed without state support. In the field of labour, the state is needed, as noted above, to pass legislation on a shorter work week, organise the redistribution of labour in society, introduce a universal basic income, job guarantees, and retraining programmes, and provide essential services to everybody. Beyond this, the state is also needed to 'take controversial decisions targeted at the power resources of affluent and influential groups' (Koch, 2024, 6).

Yet it is evident that not all states have the capacity to implement such policies in equal measure. Deva and Anand (2024) stress that due to histories of colonialism, imperialism, and neocolonialism, most Global South countries would be unable to support a universal basic income, retraining of workers, or investment in the care sector and other basic services. They suggest that what is needed for a successful sustainable transition is a global mechanism for Global North countries and businesses to 'account for historical wrongs of slavery, exploitation, and expropriation that contributed to the current level of economic development' (Deva and Anand, 2024, 107). The decolonial call to redress past injustices necessitates a global response. Relatedly, existing dependencies in the global system (e.g. around critical raw materials for batteries for electric vehicles and electronics) show that it would be impossible for any one country to undertake a degrowth transition alone, and that consequently some form of global coordination and governance is necessary (Kallis, 2015).

What does this mean for the labour governance challenges outlined earlier in this chapter? To address the problem that existing labour governance is exclusionary, Carelli (2024, 72) calls for a universal and ecological labour law, which would fully account for the 'diversity of possible kinds of work.' Anyone performing an activity that generates social and environmental value would fall under the scope of labour law, which would include the strict regulation of working time, a guaranteed basic income, and safety regulations. Extended coverage of labour governance goes hand in hand with an extended notion of social dialogue and participation. Decisions around the value of work will involve identifying first what remaining resources we are working with at the planetary level, and simultaneously where these resources are located and how they are distributed at the local level, which implies a truly multilevel notion of decision-making and citizen participation (Koch, 2024). In this context, e-governance is expected to play a bigger role in the future, in linking up the different scales of governance (Koch, 2024). This, in turn, implies a need to improve access to technology and communications infrastructure for everyone, and therefore points us back to the role of the state in providing basic services.

Further suggestions to improve inclusiveness in labour governance involve explicit mechanisms 'to recognise and reward other types of meaningful work other than formal employment' (Zekić, 2024, 67). For instance, Heintz, Staab, and Turquet (2021, 481) argue that a paradigm which values the non-market forms of work which 'nurture human capabilities and protect the natural environment' must translate into state policy which rewards workers in the paid care sector (e.g. through better terms of employment for domestic workers) as well as unpaid care workers (by unlinking benefits from labour market history and prior social security contributions, and building gender-sensitive social protection systems). Building on earlier proposals for wages for housework, Temin (2025, 178) recommends wages for earthwork: 'the core idea is that the erased labor of caring for the planet ought to be acknowledged, compensated, and materially supported in social movements, public policy, and law.' This suggestion is rooted in a conception of anticolonial climate justice which underlines the need to complement the macro dimension (countries in the Global North paying reparations to the Global South) with a micro-meso dimension (restitutions owed to indigenous people and other forms of meta-industrial labour for their stewardship of the environment, and the maintenance of ecosystems and biodiversity, individually and collectively). Beyond the financial aspect, such an approach also involves the recognition of indigenous rights and sovereignty – especially land tenure – demonstrating the close interconnection between labour and other governance areas.

In terms of enforcement, suggestions target the global level specifically. Kreinin and Aigne (2022) recommend that SDG8 be replaced with a global goal called 'Economic degrowth and sustainable work.' This should include a citizen participation indicator, a social value and environmental harm indicator, a well-being indicator, and should account for society's degree of dependence on economic growth. To ensure the implementation of global goals, Carelli (2024) argues that we need genuinely global institutions with sanctioning powers, which can issue binding rules. He suggests that the ILO could be reformed in this regard, and proposes a new structure for the ILO governing body, to include representatives from the ecological movement, who would have the right to veto policy proposals that cause environmental harm, even in instances where such policies would be beneficial to labour in the short run.

Remaining issues

While degrowth has thus prompted some interesting policy recommendations which indirectly but markedly impact on our understanding and application of labour governance, significant questions remain about the feasibility of such projects and how to implement them. For example, as mentioned above, one strand of degrowth envisions a substantial role for the state in acting against the interest of the wealthy elite and in favour of the broader population. But why would the state do so? What is needed is a strong counterhegemonic movement, capable of challenging consumerism, the growth ideology, and the current mode of living, transforming the role of the state in society. The degrowth project cannot simply rely on 'an ecologically minded global middle class willing to reduce consumerism and work-addiction' (Barca, 2024), because this will not generate the broad-based consensus necessary for change.

As Barca (2024) argues, the big question is, therefore, 'who is the political subject of a degrowth transition?' Who has an interest in degrowth? Who will struggle for such a project? She maintains that this will have to include the working class, who have the greatest vested interest in subverting the capitalist system, and who are the most affected by climate change as the newest form of class war. Extreme weather conditions hit workers the hardest, either by worsening working conditions or by undermining the infrastructure upon which their livelihoods depend (Nitsche-Whitfield, 2023). The degrowth movement, while strong on policies and ideas, is perhaps less so when it comes to articulating how exactly certain social groups (as opposed to a vague 'everybody') will benefit from the transformation, and building strategic alliances across these groups (Davidson and Gavris, 2024).

Despite this shortcoming in degrowth strategy, parts of the labour movement have begun to engage with degrowth. For example, the 2023 conference 'Beyond Growth: Pathways Towards Sustainable Prosperity,' organised at the European Parliament, included the European Trade Union Institute (ETUI) as a key partner. At this conference, the General Secretary of the European Trade Union Confederation (ETUC) highlighted important synergies between the labour movement and degrowth: a shared concern about the limitations of GDP as a measure of well-being and prosperity; a common fight for reducing the unfair imbalance between workers' wages and corporate profits, as well as the wealth of the rich; and a mutual interest in reducing working hours. However, she also pointed out that many trade unions still view growth as essential for job security and struggle to move beyond this ideology, or are sceptical of the academic nature of degrowth discourse, seen as too far removed from workers' concerns (ETUI, 2023).

The General Secretary's speech also implied that for a successful counterhegemonic alliance, degrowth must provide clear answers to key questions regarding work and labour representation. What will happen to the workers in the sectors deemed 'unnecessary' and to their unions? Will the 'necessary' sectors, to be expanded under degrowth, provide good alternatives, and, if so, how will workers be able to transition into these sectors? What will happen if production is reduced to such an extent that there are no jobs to which workers can switch? How can trade unions secure improvements in pay and working conditions for their members in a degrowth scenario? (ETUI, 2023) Although not explicitly raised in the speech, it is important to note also that some jobs, though necessary for society, can be dangerous, unpleasant, or difficult. What compensation will workers receive for performing these jobs? How to ensure that these jobs are fairly distributed in society, without reinforcing gendered and racialised inequalities? These questions underscore a critical gap: while degrowth considers work in the context of outcomes for society and the environment, it insufficiently addresses the actual process of work and working conditions (who performs the work and how).

Conclusion: what makes labour governance good?

While not claiming to have developed an all-encompassing theory of good labour governance, the analysis above nonetheless enables us to highlight some features we consider essential. In contrast to the current situation, where labour governance remains exclusionary in both concept and application, good labour governance should encompass all types of work, both industrial and meta-industrial. Genuine, broader worker participation (far beyond what exists today) is integral to good labour governance. A related feature is that good labour governance should be democratic, that is, interactive, involving stakeholder co-creation of relevant objectives, and as such might differ from polity to polity and through time.

In answering the question of 'what makes labour governance good?,' it is important to consider the social and environmental value of work: does a particular type of work contribute positively to society, or does it result in environmental damage and greater social disparity? At the same time, it is not sufficient to merely look at the consequences of a particular type of work for society at large, but we must also consider the experiences of the individuals performing the work, and the quality of the work process itself. Therefore, good labour governance is about both processes and outcomes.

The analysis above has also underscored the importance of multilevel governance. Universal labour standards and decisions about the distribution of resources and the impact of work on society and the environment require global action. But the state remains crucial for redistributing income, allocating work, and facilitating transformations in the world of work – such as reducing working hours and emphasising maintenance and care. Local communities play a vital role in mediating these large-scale transformations and adapting them to the specific needs and interests of the community.

Finally, it is important to note that the counterhegemonic movement required to drive such changes in labour governance will likely alter the very meaning of labour governance, and what labour governance is for. It could be argued that under capitalism labour governance mainly serves the interests of capital in keeping labour in check and at the same time productive; but such a perspective downplays the role of labour as a social actor (Meardi and Marginson, 2014). Even under capitalism, organised labour and social movements have fought and won important battles to obtain labour rights (e.g. in terms of working hours, health and safety at work, equal pay). In a post-capitalist world, labour governance will serve a different function than it does under capitalism, though this function remains indeterminate, as it will depend on the social forces and interests at play.

References

Barca, S. 2024. *Workers of the Earth Labour, Ecology and Reproduction in the Age of Climate Change*. London: Pluto Press.

Barrientos, S., Bianchi, L. and Berman, C. 2019. 'Gender and governance of global value chains: Promoting rights of women workers', *International Labour Review*, 158, 4: 729–725.

Barrientos, S. and Smith, S. 2007. 'Do workers benefit from ethical trade? Assessing codes of labour practice in global production systems', *Third World Quarterly*, 28, 4: 713–729.

Borges Neves, R. and Krekula, C. 2023. 'Transitions into precarity at work among older men in the metal industry in Portugal and Sweden', in Burnay, N., Ogg, J., Krekula, C. and Vendramin, P. (Eds), *Older Workers and Labour Market Exclusion Processes: A Life Course Perspective*. Cham: Springer, pp. 61–76.

Brandl, B. 2025. 'Development in social dialogue and collective bargaining', in Heyes, J., Leschke, J., Newsome, K., Reich, M. and Wilkinson, A. (Eds), *Research Handbook on the Future of Work: Decent Work in a Post-COVID 19 World*. Cheltenham: Edward Elgar, pp. 212–224.

Carelli, R. 2024. 'First lines for an ecological labour law: A social Utopia for the Anthropocene', in Bueno, N., ter Haar, B. and Zekić, N. (Eds), *Labour Law Utopias: Post-Growth and Post-Productive Work Approaches*. Oxford: Oxford University Press, pp. 71–89.

Casale, G. and Fasani, M. 2012. *International Labour Standards and Guiding Principle on Labour Administration and Labour Inspection*. Geneva: ILO.

Coslovsky, S. V. 2014. 'Flying under the radar? The state and the enforcement of labor laws in Brazil', *Oxford Development Studies*, 42, 2: 169–195.

Countouris, N. and Freedland, M. 2013. 'Labour regulation and the economic crisis in Europe: Challenges, response and prospects', in Heyes, J. and Rychly, L. (Eds), *Labour Administration in Uncertain Times*. Cheltenham/Geneva: Edward Elgar/ILO, pp. 66–94.

Davidson, J. P. L. and Gavris, M. 2024. 'Towards a degrowth transition: Bringing interests back in', *New Political Economy*, 30, 1: 48–61.

Deva, S. and Anand, P. 2024. 'A global south perspective on labour rights and supply chains for a post-growth world', in Bueno, N., ter Haar, B. and Zekić, N. (Eds), *Labour Law Utopias: Post-Growth and Post-Productive Work Approaches*. Oxford: Oxford University Press, pp. 90–115.

Durán, G. and Narbona, K. 2021. 'Precarising formality: Understanding current labour developments in Chile', *Global Labour Journal*, 12, 3: 206–226.
Durand, C., Hofferberth, E., Schmelzer, M. 2024. 'Planning beyond growth: The case for economic democracy within ecological limits', *Journal of Cleaner Production*, 437: 1–9.
ETUI. 2023. 'Trade unions 'beyond growth' – what next?', online, available from https://www.etui.org/news/trade-unions-beyond-growth-what-next, accessed 7 March 2025.
Eurofound. 2019. *Casual Work: Characteristics and Implications, New Forms of Employment Series*. Luxembourg: Publications Office of the European Union.
Fransen, L. and Burgoon, B. 2015. 'Global labour-standards advocacy by European Civil Society Organizations: Trends and developments', *British Journal of Industrial Relations*, 53, 2: 204–230.
Gavris, M. and Heyes, J. 2019. 'Varieties of labour administration in Europe and the consequences of the great recession', *Economic & Industrial Democracy*, 42, 4: 1282–1304.
Gereffi, G. and Lee, J. 2016. 'Economic and social upgrading in global value chains and industrial clusters: why governance matters', *Journal of Business Ethics*, 133: 25–38.
Graeber, D. 2018. *Bullshit Jobs: A Theory*. London: Allen Lane.
Heintz, J., Staab, S. and Turquet, L. 2021. 'Don't let another crisis go to waste: The COVID-19 pandemic and the imperative for a paradigm shift', *Feminist Economics*, 27, 1–2: 470–485.
Hickel, J. 2020. *Less is More: How Degrowth Will Save the World*. London: Penguin.
Hickel, J. and Kallis, G. 2020. 'Is green growth possible?', *New Political Economy*, 25, 4: 469–486.
Hyman, R. 2013. 'The role of government in industrial relations', in Heyes, J. and Rychly, L. (Eds), *Labour Administration in Uncertain Times*. Cheltenham/Geneva: Edward Elgar/ILO, pp. 95–124.
ILO. 2011. Report V - Labour Administration and Labour Inspection. *International Labour Conference 100th Session*. Geneva: ILO.
ILO. 2016. *Non-Standard Employment Around the World: Understanding Challenges, Shaping Prospects*. Geneva: ILO.
ILO. 2021. *Making Decent Work a Reality for Domestic Workers: Progress and Prospects ten Years after the Adoption of the Domestic Workers Convention*, 2011 (No. 189). Geneva: ILO.
ILO. 2023. *Women and Men in the Informal Economy: A Statistical Update*. Geneva: ILO.
ILO. 2024. *Social Dialogue Report 2024: Peak-Level Social Dialogue for Economic Development and Social Progress*. Geneva: ILO.
Jaehrling, K. and Méhaut, P. 2013. 'Varieties of institutional avoidance': Employers' strategies in low-waged service sector occupations in France and Germany', *Socio-Economic Review*, 11: 687–710.
Kallis, G. 2015. 'The degrowth alternative', online, available from https://greattransition.org/images/Kallis-Degrowth-Alternative.pdf, accessed 7 March 2025.
Kreinin, H. and Aigne, E. 2022. 'From "Decent work and economic growth" to "Sustainable work and economic degrowth": A new framework for SDG 8', *Empirica*, 49: 281–311.

Krings, T. 2014. Varieties of social dumping in an open labour market: The Irish experience of large-scale immigration and the regulation of employment standards. *ETUI Policy Brief*, No 6/2014. Brussels: ETUI.

Koch, M. 2024. 'Deepening the degrowth planning debate: Division of labor, complexity, and the roles of markets and digital tools', *Sustainability: Science, Practice and Policy*, 20, 1: 1–13.

LeBaron, G. 2020. *Combatting Modern Slavery: Why Labour Governance Is Failing and What We Can Do about It*. Cambridge: Polity Press.

Li, M. and Hu, X. 2022. 'Can labor-standards advocacy by transnational civil society organizations interact with the power of labor to improve labor standards in global supply chains? A case study of the Cambodian garment industry', *Economic & Industrial Democracy*, 44(1), 208–229.

Locke, R. 2013. *The Promise and Limits of Private Power Promoting Labor Standards in a Global Economy*. New York: Cambridge University Press.

Locke, R., Amengual, M. and Mangla, A. 2009. 'Virtue out of necessity? Compliance, commitment and the improvement of labor conditions in global supply chains", *Politics & Society*, 37, 3: 319–351.

Locke, R., Rissing, B. and Pal, T. 2013. 'Complements or substitutes? Public vs. private regulation and the enforcement of labor standards in global supply chains", *British Journal of Industrial Relations*, 51, 3: 519–552.

Lucassen, J. 2021. *The Story of Work: A New History of Humankind*. New Haven, CT: Yale University Press.

Marginson, P. and Sisson, K. 2004. *European Integration and Industrial Relations: Multi-Level Governance in the Making*. Basingstoke: Palgrave MacMillan.

Marx, A. and Wouters, J. 2016. 'Redesigning enforcement in private labour regulation: Will it work?', *International Labour Review*, 155, 3: 435–459.

Meardi, G. and Marginson, P. 2014. 'Global labour governance: Potential and limits of an emerging perspective', *Work, Employment & Society*, 28, 4: 651–662.

Nelson, A. 2024. 'Marx, work, agency and postcapitalist prefiguration'. *Human Geography*, online before print, https://doi.org/10.1177/19427786241237434, pp 1–11.

Nitsche-Whitfield, P. 2023. 'Beyond economic growth The role of trade unions in the transition to well-being', ETUI report, online, available from https://www.etui.org/publications/beyond-economic-growth#:~:text=It%20is%20widely%20accepted%20that,on%20biodiversity%20is%20ecologically%20unsustainable, accessed 12 March 2025.

OECD/ILO. 2019. *Tackling Vulnerability in the Informal Economy, Development Centre Studies*. Paris: OECD Publishing.

Papadakis, K. 2021. 'Social dialogue at the dawn of the ILO's centenary: Sorting out challenges, setting priorities for the future', in Heyes, J. and Rychly, L. (Eds), *The Governance of Labour Administration: Reforms, Innovations and Challenges*. Cheltenham/Geneva: Edward Elgar/ILO, pp. 43–67.

Pérez Orozco, A. 2022. 'Ecofeminist degrowth for sustaining Buen Convivir', *Hypatia*, 37: 223–240.

Pettifor, A. 2019. *The Case for the Green New Deal*. London: Verso.

Rogan, M., Roever, S., Alter Chen, M. and Carré, F. 2018. 'Informal employment in the global south: Globalization, production relations, and "precarity"', in Kalleberg, A. L. and Vallas, S. P. (Eds), *Precarious Work: Research in the Sociology of Work*. Bingley: Emerald, Vol. 31: 307–333.

Rossi, A. 2015. 'Better work: Harnessing incentives and influencing policy to strengthen labour standards compliance in global production networks', *Cambridge Journal of Regions, Economy and Society*, 8: 505–520.
Saave, A. and Muraca, B. 2021. 'Rethinking labour/work in a degrowth society', in Räthzel, N., Stevis, D. and Uzzell, D. (Eds), *The Palgrave Handbook of Environmental Labour Studies*. London: Palgrave Macmillan, pp. 743–767.
Salleh, A. 1997. *Ecofeminism as Politics*. London: Zed Books.
Srivastava, R. 2016. Structural change and non-standard forms of employment in India. *Conditions of Work and Employment Series* No. 68. Geneva: ILO.
Temin, D.M. 2025. 'Wages for Earthwork'. *American Political Science Review*, 119(1), 179–192: https://www.cambridge.org/core/journals/american-political-science-review/article/wages-for-earthwork/BD4D0D6F84389EB65153E9C19E42F7CF
Thörnquist, A. 2013. 'False (bogus) self-employment in East-West labour migration: Recent trends in the Swedish construction and road haulage industries', *Themes on Migration and Ethnic Studies*, 41.
Vincent, O. and Brandellero, A. 2023. Transforming work: A critical literature review on degrowth, post-growth, postcapitalism and craft labor, *Journal of Cleaner Production*, 430: 1–9.
Wagner, I. and Bernsten, L. 2016. 'Restricted rights: Obstacles in enforcing the labour rights of mobile EU workers in the German and Dutch construction sector', *Transfer*, 22, 2: 193–206.
Weil, D. 2014. *The Fissured Workplace: Why Work Became So Bad for So Many and What Can Be Done to Improve It*. Cambridge, MA: Harvard University Press.
Zekić, N. 2024. 'Labour law for degrowth and meaningful work', in Bueno, N., ter Haar, B. and Zekić, N. (Eds), *Labour Law Utopias: Post-Growth and Post-Productive Work Approaches*. Oxford: Oxford University Press, pp. 51–70.

4

GENDER EQUALITY AND GOOD GOVERNANCE

Women's political leadership and participation

Marta Guerriero

Introduction

This chapter aims to explore the often-understudied relationship between good governance and gender equality.[1] Traditionally, the study of gender inequality has been confined to the *private* sphere of domestic life and work, as opposed to the *public* realm of governance and politics (Nussbaum, 2003). However, growing global awareness in recent years that gender equality is not only a basic human right but also essential to the achievement of sustainable development (OECD, 2019; Leal Filho et al., 2023) has brought the study of the relationship between governance and gender to the attention of sustainable development scholars.

Despite the recent popularity of approaches aimed at promoting gender equality among many international agencies and national governments, gender inequality persists, to varying degrees, around the world and in different contexts (UN Women, 2025). A significant area is women's political leadership and participation. From the local to the global level, women are often excluded from decision-making, and they face several obstacles to participation and leadership in political and public life (Inter-Parliamentary Union, 2025). Although the global agenda on good governance highlights the significance of issues such as participation, equity, and accountability, it does not offer the necessary attention to women's inclusion in politics.

By focusing on women's political leadership and participation as a case study, I argue that the relationship between gender equality and good governance is bi-directional: good governance contributes to gender equality in politics and, in turn, sustainable development; simultaneously, gender equality in political and governance structures is key to achieving inclusive

DOI: 10.4324/9781003467694-5

and accountable governance, and promoting several, if not all, Sustainable Development Goals. I also critique mainstream conceptualisations and operationalisations of good governance for their lack of explicit reference to both women and intersectionality, contending that overlapping identities – such as race, class, ethnicity, and sexuality – shape women's access to political power and governance outcomes. Governance accountability should therefore be assessed in both gender and intersectional terms.

The remainder of this chapter is organised as follows. The following two sections explain the relationship between gender equality and sustainable development, whereby gender equality is both a goal and cross-cutting issue in the Sustainable Development Agenda, and a precondition for sustainable development. I advocate for the need to shift the narrative on women and development towards a greater acknowledgment of the role of women as *agents* of sustainable change. A further section then presents the main idea of a dialectical relationship between gender equality and good governance, and illustrates this with cross-country data on governance and gender inequality. This is followed by a concrete example, that of women's political leadership and participation, which is used to highlight both the agential role of women and the mutual relationship between gender and governance. Some of the academic literature on gender equality and good governance is then critiqued, by emphasising the need to focus on rights-based gender equality approaches (as opposed to utility-based approaches) and an intersectional lens (as opposed to a mere gender lens). Finally, this chapter ends with a critique of mainstream definitions and measures of good governance, followed by a summary and conclusions.

Gender equality as a goal and cross-cutting issue in the Sustainable Development Agenda

Recent decades have led to growing global recognition that gender equality is not only a basic human right, but also a fundamental goal of sustainable development (OECD, 2019), with gender equality becoming a cross-cutting issue in the implementation of the Sustainable Development Agenda (Leal Filho et al., 2023). Scholars have described recent times as being characterised by a 'rising tide of support for gender equality,' with the pursuit of gender equality now widely endorsed as a central policy goal, both nationally and internationally (Inglehart & Norris, 2003, 10).

The period from 1975 to 1995, in particular, marked a number of landmark global agreements on women's rights, including the Convention on the Elimination of All Forms of Discrimination against Women (CEDAW) in 1979, the Cairo International Conference on Population and Development (ICPD) in 1994, and four United Nations (UN) World Conferences on Women (Mexico City 1975, Copenhagen 1980, Nairobi 1985, Beijing 1995), leading to an explicit Platform for Action for policymakers ensuring that a gender perspective was reflected in policies and programmes at all levels (Zinsser, 2002; Tarr-Whelan, 2010). This gave rise to the global explosion of Women's Policy Agencies (WPAs), government bodies aimed at enhancing policy development and implementation around women's specific concerns, with the potential – according to the 1995 Beijing Platform for Action – 'to support government-wide mainstreaming of a gender-equality perspective in all policy areas' (United Nations, 1995, 84). By the mid-1990s, 127 countries across the globe had created such offices (McBride & Mazur, 2013). Gender mainstreaming strategies, incorporating a gender perspective into all areas of policy, became popular in the discourse of many international agencies and national governments and, despite being also highly debated (Bacchi & Eveline, 2010), they have been described as 'an international phenomenon' (Walby, 2005, 453).

Although the start of the Millennium was often seen as a hold up on progress on women's rights and gender equality, with the Millennium Development Goals (MDGs) commonly viewed as a setback for rights-based approaches to development (Kabeer, 2015), the Sustainable Development Goals (SDGs) introduced in 2015 offered renewed optimism. In particular, scholars praised their two-track strategy: a stand-alone goal dedicated to a broad range of issues related to gender equality and women's empowerment (SDG 5) and explicit articulation of gender matters in a number of other goals (Esquivel & Sweetman, 2016), highlighting the multidimensional and cross-cutting nature of gender equality.

Nevertheless, gender equality itself is a contested idea, discussed among both theorists and practitioners who propose a variety of visions and debates in relation to it (Verloo & Lombardo, 2007). Furthermore, despite having seen global advances in gender equality in a number of different areas, the progress is still slow (Leal Filho et al., 2023). Notwithstanding the attempts of governments and multilateral organisations to increasingly introduce gender equality strategies and policies, these efforts often fall short, not supported by effective governance strategies.

For example, in relation to the economic pillar of sustainable development, women are more likely to experience poverty than men (UN Women, 2025). Globally, the labour force participation rate[2] is only 48.5% for women, compared to 74.1% for men (International Labour Organization, 2024), with a significant number of working-age women neither working nor actively looking for work. Indeed, women devote a considerably greater amount of time than men (approximately 2.5 times as many hours per day) to unpaid care and domestic work (UN Women, 2025). When considering more social dimensions of sustainable development, the picture is similar. Women are more likely than men to experience sexual violence, and only 56.3% of adult[3] married (or in a union) women worldwide have decision-making power over their sexual and reproductive health and rights (UN Women, 2025). Furthermore, globally, 70.2% of adult[4] men have at least some secondary education, contrary to 62% of adult women (UNDP, 2025). Finally, with respect to the environmental pillar of sustainable development, there is considerable evidence that women are more likely to die from natural disasters than men (OECD, 2025). Moreover, it is predicted that approximately 75% of the millions of people at risk of displacement due to climate change and severe environmental hazards will be women (OECD, 2021), a significant overrepresentation of the female population.

Gender equality as a precondition for sustainable development: from women as *vulnerable groups* to women as *agents of change*

Too often, however, in discussions around the relationship between gender and sustainable development, the literature focuses on women as vulnerable groups (Huyer et al., 2020), portraying women as *victims* of climate change, capitalism and/or traditional development models and policies, which have contributed to reinforcing inequality, gender relations and roles (Leach et al., 2016). This is particularly the case when gender intersects with characteristics such as socio-economic status, ethnicity, age and other significant axes of difference. In fact, according to the intersectionality framework (Crenshaw, 1989), in order to understand women's experiences of oppression and marginalisation, it is important to appreciate how gender differences and roles interact with and exacerbate other forms of discrimination, in both the private and the public spheres.[5]

Nonetheless, such focus on vulnerability has not always led to successful policymaking (Stevens, 2010). As mentioned in the previous section, despite global efforts, gender inequality persists, to varying degrees, around the world and in different contexts. Additionally, the growth of recent movements such as #MeToo (a global awareness movement on sexual abuse and sexual harassment) and #HeforShe (a UN's solidarity campaign on the advancement of gender equality) further demonstrates the public opinion's desire and urgency to see greater progress in gender equality (Collier Hillstrom, 2019; Harvey, 2020) and to put women at the forefront of global decision-making. It is therefore imperative to shift focus, and in particular to recognise that gender equality can significantly benefit sustainable development (Leach et al., 2016). Women's knowledge, agency and collective action should be seen as central – according to Candice Stevens (2010, 1), they are 'key' – to achieving sustainable development.

A growing body of academic literature suggests that gender equality and women's empowerment can deliver positive impacts on sustainable development and the 2030 Agenda, triggering positive multiplier effects on a number of Sustainable Development Goals (Dugarova, 2018). Women are fundamental in managing natural resources, producing and accessing food, and a number of other key services such as education, health, sanitation, and energy. An analysis conducted by the United Nations Development Programme on the economic impacts of investing in gender equality in education and labour force participation found that equal educational attainment between men and women and a 50% increase in the female share of the total labour force by 2030 would contribute US$4.4 trillion to global Gross Domestic Product (Dugarova, 2018). Similarly, estimates by the Food and Agriculture Organization indicated that reaching gender-equal access to productive resources and opportunities could increase crop yields by 20–30% and lift 100–150 million people out of hunger (FAO, 2011). Furthermore, due to their unique understanding of nature and experience over the management of natural resources, women are stewards of the environment and key actors in climate change mitigation and adaptation (Dankelman & Fansen, 2010).

Because of the important implications that gender equality has on all three pillars of sustainable development – People, Planet, and Prosperity – a fundamental shift must be advocated: from more traditional models focusing on women as *victims* of uneven and unsustainable development, towards more modern approaches that place women as central *agents* of sustainable change. In 1992, at the Rio Earth Summit, the UN called for women to be more actively involved in sustainable development decision-making. Although we are regularly reminded of this call (Buckingham, 2010; OECD, 2021), women still do not feature as prominently as they should in sustainable development discussions. Consequently, the slow advancements on gender equality described in the previous section may be a primary cause of the failure to also make significant global progress on sustainable development.

In summary, gender equality is not only a fundamental goal of global sustainable development, but it is also an essential precondition for its achievement. On the one hand, progress on sustainable development has positive implications on the conditions of women and girls. On the other hand, gender equality is instrumental to achieving sustainable development. Applying a gender equality lens when addressing issues of sustainable development is therefore essential, as gender equality and sustainable development can help to reinforce each other.

The role of good governance in the gender-sustainable development nexus: mutually reinforcing mechanisms

The study of the relationship between gender and governance is relatively recent in the academic literature, since traditional work used to distinguish between the *public* sphere of governance and politics and the *private* realm of domestic life and work, where gender inequalities were expected to arise (Nussbaum, 2003; Waylen, 2008). Sharon Bessell (2016, 284) described this separation very effectively, as 'a disjuncture [...] between the global good governance agenda and the global gender equality agenda'.

However, as seen in the previous sections, recent work on gender inequality has recognised its multidimensionality and cross-cutting nature (Leal Filho et al., 2023), including its significant public and governance dimensions. Therefore, when exploring the above-described nexus between gender and sustainable development, it is impossible not to consider how it interplays with governance. Given its multiple ramifications on all three pillars of sustainable development, gender equality in governance structures is often described as key to promoting several, if not all, SDGs. Consequently, gender inequality is not only a fundamental obstacle to achieving sustainable development, but also a barrier to good governance. In turn, good governance – a central pillar of sustainable development itself – may contribute to gender equality and, as a result, sustainable development. As a matter of fact, the international agreements and legislations on gender equality mentioned above require translation into national and local policies that need to be implemented on the ground via legal directives, financial budgets, institutional arrangements, bureaucratic procedures, monitoring, etc. In other words, they require governance practices and processes (UNRISD, 2005).

Good governance and gender equality are therefore expected to be mutually reinforcing. Governance that is transparent, accountable and inclusive will create spaces for women's empowerment, while gender equality will strengthen governance by making institutions more representative and just. In the absence of gender equality, governance cannot be defined as 'good'; simultaneously, gender equality cannot be achieved without good governance.

Figure 4.1 illustrates the mutually reinforcing relationship between good governance and gender inequality. The six charts are scatter diagrams showing linear correlations between the Gender Inequality Index (GII), on the y-axes, and the six Worldwide Governance Indicators (WGIs), on the x-axes. GII is a composite metric of gender inequality published by the United Nations Development Programme (UNDP, 2025) and constructed by combining three dimensions: health,[6] empowerment,[7] and labour market.[8] The index ranges from 0 to 1, with lower GII scores corresponding to lower gender inequality. The WGIs, created by the World Bank, are aggregate indicators of six dimensions of governance: voice and accountability, political stability and absence of violence/terrorism, government effectiveness, regulatory quality, rule of law, and control of corruption (World Bank, 2024). All indicators are measured as estimates of governance in standard normal units, ranging from −2.5 to +2.5, where a higher score indicates stronger governance performance. The data in Figure 4.1 refers to 2023 (the most recent available year for both datasets) and 172 countries (the ones for which we possess data on both governance and gender inequality).

Figure 4.1 The relationship between good governance and gender equality.

Source: Author's own, based on UNDP (2025) and World Bank (2024) data.

The diagrams display a strong negative correlation between good governance and gender inequality for all six indicators, whereby countries with lower gender inequality tend to also exhibit stronger governance performance. The correlation coefficient is stronger for government effectiveness (–0.81) and regulatory quality (–0.80), where the data points lie close to a line of best fit, followed by rule of law (–0.78) and control of corruption (–0.74), and it is weaker for political stability and absence of violence (–0.60) and voice and accountability (also –0.60), where the data points are significantly more dispersed.[9]

Women's political leadership and participation: a case study

One way of advancing gender equality that places women as central *agents of change* is by promoting equal participation in decision-making between men and women at a number of levels (Leach et al., 2016), including in the public and political sphere, which is a prerequisite for fair and effective

TABLE 4.1 Global and regional averages of women in national parliaments

	Single chamber or lower chamber (%)	*Upper chamber (%)*	*All chambers combined (%)*
World	27.2	27.8	27.3
Americas	35.7	36.0	35.8
Europe	31.6	32.2	31.7
Sub-Saharan Africa	26.8	27.2	26.9
Asia	21.8	22.8	21.9
Pacific	21.3	49.5	24.3
Middle East and North Africa	18.8	10.3	17.0

Source: Author's own based on data from Inter-Parliamentary Union (2025).

governance. Public and political institutions that better reflect the diversity of the population are indeed expected to contribute to fairer and more equitable decision-making, improved public trust, and policy priorities that reproduce a broader consensus (UN Women, 2025).

Nevertheless, from the local to the global dimension, women are often excluded from decision-making, and they face several obstacles to participation and leadership in the political and public life. Globally, as of the 1st October 2025, women hold 27.2% of seats in national parliaments, with only seven countries in the world with 50% or more women in the lower or single chambers (Inter-Parliamentary Union, 2025). 102 countries have never had a woman Head of State or Government (UN Women, 2025). As presented in Table 4.1, according to the IPU (Inter-Parliamentary Union, 2025), the proportion of women in the lower or single chambers varies significantly across regions in the world,[10] and it is highest in the Americas (35.7%) and Europe (31.6%), and lowest in the Middle East and North Africa (18.8%), Asia (21.8%), and the Pacific (21.3%).

Women's representation in local governments is also low, at 35.5%, and the representation ratio in public service (compared to population shares) also falls short of gender parity, at 0.80. The gap is especially pronounced in Central and Southern Asia, at 0.52 and 0.46 respectively. Furthermore, data disaggregated by public service position show that women are disproportionately concentrated in clerical and administrative roles, remaining underrepresented in senior government positions (0.74) (UN Women, 2025).

Low representation of women in political leadership brings fundamental concerns in terms of the laws and policies currently being drafted, and it forces us to question the quality of existing laws and policies, not just in relation to how they have been promoting gender equality (Bessell, 2016). Indeed, there is empirical evidence from a range of countries that women's leadership in political decision-making influences political outcomes, especially in favour of health, education and social welfare. Chattopadhyay and Duflo (2004)'s seminal research on Indian *panchayats* (local councils) found

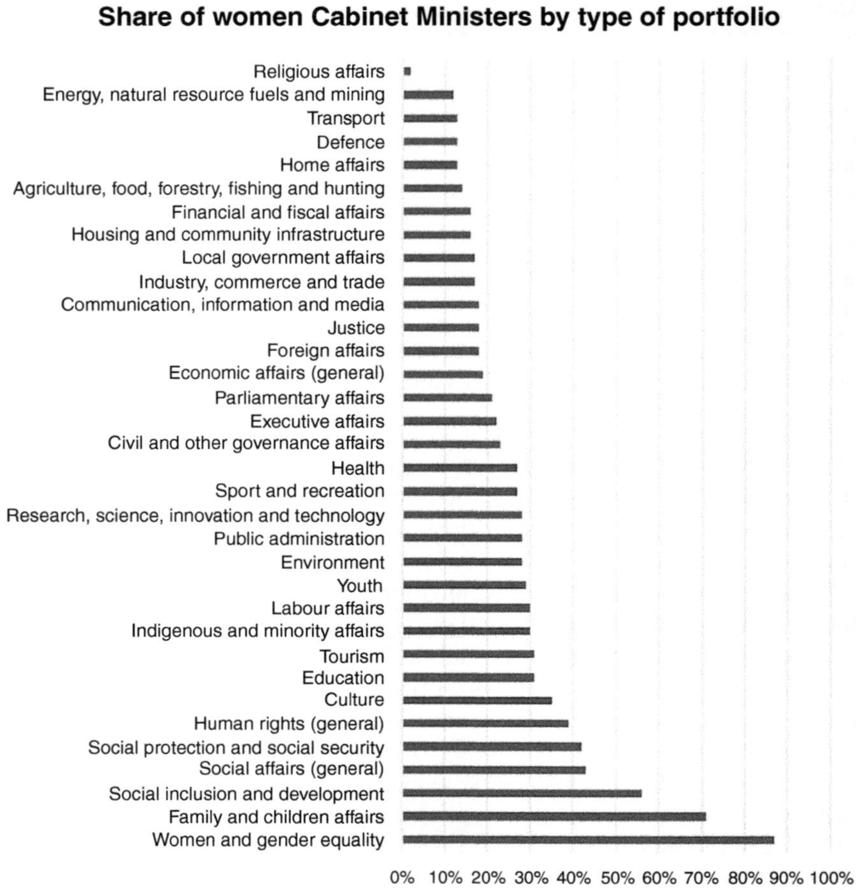

Figure 4.2 Share of women Cabinet Ministers by type of portfolio.

Source: Author's own, based on data from Inter-Parliamentary Union & UN Women (2025).

that women political leaders tend to invest more in public goods such as drinking water. In wealthier countries, there is evidence that women in politics focus more on legislation and policies in areas such as parental leave and child support, gender-based violence and other issues that feel more relevant to women. In Norway, for example, Bratton and Ray (2002) observed a relationship between female representation in municipal councils and childcare coverage. Although the above empirical studies are two decades old, they have been influencing empirical research on the impacts of women as policymakers to this day. For example, Baskaran and Hessami (2025) found that female councillors accelerate the expansion of public childcare in Bavaria. Similar evidence can also be seen by exploring current data on the share of women Cabinet Ministers by type of portfolio[11] (Inter-Parliamentary Union & UN Women, 2025) displayed in Figure 4.2. The categories of portfolios that observe greater representation of women globally are human rights, gender equality and social affairs, education, and environment. These align with the areas of political interest mentioned above: health, education, gender and social welfare. On the contrary, the portfolios where women are least represented are economic affairs, housing and infrastructure, justice and home affairs, and defence.

Further academic literature on gender equality and good governance highlights a relationship between gender equality and democratic norms (Cornwall & Goetz, 2005), as well as a relationship between female political representation and engagement with corruption and clientelism (Goetz, 2007; Bryant et al., 2025), with women parliamentarians found to be introducing different values, issue priorities and political behaviours compared to their male counterparts (Squires, 2007).

The vast empirical evidence on the relationship between women's political participation and good governance stimulated the introduction of measures such as gender quotas (or reservations) as a way to secure increased numbers of women in policymaking and promote gender equality in political participation and leadership. Balanced political participation and power-sharing in decision-making between women and men is also an internationally agreed target set in the Beijing Declaration and Platform for Action (United Nations, 1995). Gender quotas, which were first introduced in Western Europe in the 1970s within political parties, have now been included in the legislation at the parliamentary level in 84 countries globally (44% of the world), according to the Economic and Social Council (ECOSOC, 2021). While most countries in the world have not achieved gender parity, gender quotas have certainly helped to drive improvements over the years. For example, women's seats in national parliaments have seen a 4.9 percentage point increase globally since 2015, and the number of countries with 50% or more women in the lower or single chambers has also doubled in this period. In countries with legislated candidate quotas, women's representation is five percentage points higher in parliaments and seven percentage points higher in local governments respectively, compared to countries without such legislation (ECOSOC, 2021). With 63.8% women in the lower chamber and 50% women in the upper chamber, Rwanda has one of the highest figures in the world for women's representation in parliament (Inter-Parliamentary Union, 2025). This is in large part due to a well-established gender quota system in the country. Overall, gender quotas have had a positive, but mixed, impact on women's representation, being most effective in proportional representation electoral systems, when political parties are supportive (Squires, 2007).

Besides gender quotas, another key strategy aimed at promoting women's inclusion in policymaking and the political sphere is gender mainstreaming.

As defined by the Economic and Social Council 1997 'Agreed Conclusions on gender mainstreaming' (ECOSOC, 1997, 3),

> Mainstreaming a gender perspective is the process of assessing the implications for women and men of any planned action, including legislation, policies or programmes, in all areas and at all levels. It is a strategy for making women's as well as men's concerns and experiences an integral dimension of the design, implementation, monitoring and evaluation of policies and programmes in all political, economic and societal spheres so that women and men benefit equally and inequality is not perpetuated. The ultimate goal is to achieve gender equality.

Gender mainstreaming calls for policymakers to take gender into account when writing, implementing and evaluating legislation, but it does not necessarily require policymakers to be themselves women, as advocated by gender quotas. The impact of gender mainstreaming on women's political representation has proven difficult to establish, often associated with the inclusion of gender impact assessments into the policymaking process and the expansion of the range of actors involved (Squires, 2007).

Gender equality: intrinsic or instrumental to good governance?

The above-mentioned literature and approaches advocate for gender equality in politics to be pursued because it is an instrument for improved governance practices. The case for increased participation of women in politics and public service is fundamentally made on efficiency grounds (King & Mason, 2001). Nevertheless, gender equality should not be perceived only in utility-based terms, as an *instrument* for good governance. Intrinsic and rights-based motivations of social justice should also justify the inclusion of gender-sensitive practices in political governance and these – as argued by some scholars (Squires, 2007) – are fundamental. Women's political participation and leadership have both a symbolic and a substantive value, as they contribute to challenging patriarchal norms and addressing systemic discrimination, along with influencing policy outcomes. As Janet Halley et al. (2018, x) argue, *Governance Feminism* has indeed led to 'immense emancipatory achievements'. Rights-based approaches should therefore not be overlooked when designing appropriate policies and practices, and they may also help us to re-politicise gender equality, as mainstream methods have been critiqued for not being sufficiently driven by intrinsic motivations (Cornwall & Molyneux, 2006; Cornwall & Nyamu-Musembi, 2005). As a matter of fact, both gender quotas and gender mainstreaming have been widely debated: quotas have generated extensive discussions on the 'politics of presence' (Phillips, 1998, 5), while mainstreaming has caused debate on the 'politics of process' (Squires, 2007, 3). Both strategies have been criticised for adopting formal empirical indicators – the former is centred on numerical indicators, the latter focuses on gender experts and implementation processes such as gender impact assessments – which render gender equality *measurable* but more distant from its intrinsic meaning and original conceptualisation. As mentioned in the previous section, women's political leadership should be recognised as central in redefining governance priorities. Women's representation should be substantive, not just numerical, with women influencing policy outcomes rather than merely occupying seats. Gender quotas have certainly facilitated women's entry into politics, but at a relatively superficial level, without tackling the root causes of unequal access to politics (Nussbaum, 2003). Making governance gender-sensitive requires more than adding women in parliaments (in other words, women's *presence*) – it requires them building sufficient capacity. Concerning gender mainstreaming, efforts towards gender equality should not just be linked to institutional and technocratic practices (in other words, *processes*), but they should be based on rights-based gender equality arguments and participatory approaches. In summary, scholars argue that we should stop focussing on *presence* and *process*, and that we should return to an emphasis on *voice* (Squires, 2007).

Furthermore, although it is important to view women as *agents* of good governance, we should ensure that we do not instrumentalise gender equality by over-relying on the agential capacity of women – individually or collectively – to bring about governance reform and sustainable change. Firstly, this risks placing a disproportionate burden on women, rather than on societies and governments (Bell & Wegner, 2025). Secondly, this also misses critical structural and intersectional considerations, and helps to explain why the above efforts have had limited impact on gender equality. Gender-sensitive governance requires fundamental changes to reorientate public action and policy, and it needs to start from the recognition that the institutions and processes of governance are gendered and therefore need to be challenged.

Gender equality and good governance: the importance of an intersectional lens

Contrary to the academic and policy literature mentioned in an earlier section, on women's vulnerability and marginalisation, which often incorporates considerations of intersectionality, research on gender and good governance has traditionally studied *women* as a single overarching category in both analysis and policy. Although this may be due to limited data availability on the intersection between gender and other demographic characteristics – such as race, class, ethnicity, disability and sexuality –, there are critical social differences within the category *women* that should not be overlooked (Crenshaw, 1989). Relatively recent research on gender and politics highlights the critical importance of considering intersectionality when discussing women's political participation and leadership (Hancock, 2007). Celis and Childs

(2020; 2023) call for a *second generation* of feminist scholarship aimed at foregrounding intersectional heterogeneity. Christoffersen (2021) highlights the need to focus on diversity among women, but also acknowledges the difficulties of conceptualising and operationalising intersectionality in policy and practice. Christoffersen and Siow (2025, 671) emphasise the need to focus on 'intersectionally marginalised constituencies' in particular.

Accordingly, policies and interventions aimed at the inclusion of women in politics that are insensitive to differences *between* women (or *within* women, when these are interpreted as a uniform category) might be counterproductive. For example, unless appropriately designed, policies aimed at women's representation within political leadership (aimed at *de jure* equality) may lead to elite women gaining access to political office and capturing the benefits, leaving marginalised and less advantaged groups of women excluded (and therefore still leading to *de facto* inequality and stratification, within women instead of between women and men). A study by Celis et al. (2014) comparing Belgium (where legally binding quotas exist) and the Netherlands (where they do not exist) found that gender quotas as a form of equality policy might negatively affect ethnic minority groups. Additionally, when examining gender-based quotas for candidates in legislative elections in Brazil, Tosi (2025) found that, for female candidates especially, characteristics like race, education and marriage status significantly increase their likelihood of being elected. Finally, it has been observed that governance frameworks that are intersectionality-neutral may lead to policy *blind spots*, disproportionately impacting marginalised women. For example, there is no specific reference to Black women and girls in current trafficking policy and legislation in the United States (Dodd & Schroeder, 2024), a policy area significantly influenced by women's political leadership (Perry & Burns, 2023), despite the fact that Black women and girls are the most common victims of sex trafficking in the country.

How to measure good governance in relation to gender and intersectionality?

Despite highlighting significant issues such as participation, equity and accountability, the global agenda on good governance does not seem to offer the necessary attention to women's inclusion. The above-mentioned World Bank framework (World Bank, 2024) in particular, which has been shaping global understandings of good governance, does not sufficiently consider women's inclusion in governance processes. While it acknowledges gender equality in action plans, its core definition of good governance is gender-neutral. The governance reforms advocated by the World Bank's technocratic approach are concerned with the effectiveness, efficiency and accountability of institutions, and not with issues of participation, human rights and social justice, and do not take into account how institutions may be shaped by unequal gender relations and may reproduce gender divisions (UNRISD, 2005). By failing to focus on women, and in particular women's political participation and leadership, the World Bank's framework risks treating governance as a technical matter rather than a political one, and consequently perpetuating governance models that may be efficient but not equitable.

The WGI framework (World Bank, 2024) introduced above does not explicitly measure women's political participation and it assumes a universal citizen, ignoring both gender and how other characteristics intersect with gender to shape governance outcomes. None of the indicators incorporate any consideration of gender, nor is there any reference to gender in the background paper (Bessell, 2016). Looking at a specific indicator, for instance 'Voice and Accountability' (World Bank, 2024), which in Figure 4.1 is not as strongly associated with gender inequality as others, this measures *voice* in abstract and universal terms. However, we know from gender and intersectionality theory that not all voices are equally heard, with women – and women from minority groups in particular – often facing exclusion. Overall, quantitative indicators such as the WGIs tend to ignore individual specificity and context in favour of standardised knowledge (Engle Merry, 2011), missing out important gender- and intersectionality-sensitive information and risking to further perpetuate the marginalisation of women and, in particular, disadvantaged groups of women. Other critiques of mainstream definitions of good governance highlight the fact that the governance reforms proposed by institutions like the World Bank privilege Western models of both democracy and feminism (Leftwich, 1993; Thomas, 2010), universalising women and overlooking intersectionality – and therefore potentially reproducing global inequalities among women – but also excluding indigenous and local forms of leadership and feminism in the Global South (Mohanty, 2003).

Consequently, indicators of good governance should better capture and reflect the differential impacts of and participation in the mechanisms and processes of governance between men and women (i.e. they should be gender-sensitive), and also within women (i.e. they should be intersectional and pro-poor) (Corner, 2005). A true conceptualisation and operationalisation of good governance should explicitly include women in its indicators, integrate gender equality with intersectionality, and disaggregate data on women by race, class and other identities. Governance accountability should be assessed in both gender and intersectional terms, by monitoring how it addresses inequalities across multiple axes of identity.

Conclusions

By focusing on women's leadership and participation, this chapter explains and evidences the presence of a nexus between good governance and gender equality, and its implications on sustainable development. The relationship between gender and governance is mutually reinforcing, with good governance contributing to gender equality, gender equality being key to achieving inclusive and accountable governance, and both gender equality and good governance being fundamental to promoting sustainable development.

Good governance must be gender-sensitive and offer the necessary attention to women. However, by doing so, it should not treat women as a single uniform category, but it should acknowledge fundamental differences among women and how gender intersects with other critical social characteristics. Moreover, gender-sensitive governance requires the realisation of women's rights to be central to the aims and practices of governance itself, as achieving greater gender equality in governance is an intrinsically important end.

Until today, the global agenda on good governance has not offered the necessary attention to women's inclusion, with mainstream conceptualisations and operationalisations of governance failing to make explicit reference to both women and intersectionality. Genuinely good governance should integrate gender equality and intersectionality as core principles and recognise that gender- and intersectionality-inclusive political leadership and participation is not optional but essential for sustainable development. Applying these lenses to the different processes and levels of decision-making, and placing greater emphasis on the gendered and intersectional nature of citizenship and power structures is required to build better governance and, in turn, generate concrete benefits for People, Planet, and Prosperity.

In summary, good governance strategies should be intersectional and pro-poor, they should support diverse women's agency, power and voice in political institutions and decision-making, they should build political forms that target the Sustainable Development Goals in gender- and intersectionality-inclusive ways, and they should create spaces for women's diverse voices and political actions.

Notes

1 Although we acknowledge that 'gender' is a contested term and that it is a non-binary construct, when referring to 'gender' in this chapter we focus on women as opposed to men. We intentionally refer to 'gender' more often than 'women' to reflect the idea that societies construct definitions of what women and men are.
2 The International Labour Organization defines the labour force participation rate as the share of people in the labour force as a percentage of the entire working-age population. The labour force comprises all persons of working age who are either employed or unemployed.
3 'Adult' here is defined as aged 15–49 years.
4 'Adult' here is defined as aged 25 years and older.
5 Intersectionality in relation to good governance and gender equality is discussed in more detail later in this chapter.
6 An index on female reproductive health is constructed by combining data on maternal mortality and adolescent birth rate.
7 This is measured by bringing together data on female and male population with at least secondary education and female and male shares of parliamentary seats.
8 In other words, female and male labour force participation rates.
9 We will return to the indicator on 'Voice and Accountability' in the penultimate section of this chapter.
10 Regions are displayed in Table 4.1 in descending order of the percentage of women with seats in unicameral parliaments or in the lower chamber of the parliaments.
11 Based on data collected for 189 countries in the world.

References

Bacchi, C. & Eveline, J., 2010. *Mainstreaming Politics: Gendering Practices and Feminist Theory.* Adelaide: University of Adelaide Press.

Baskaran, T. & Hessami, Z., 2025. Women in Political Bodies as Policymakers. *The Review of Economics and Statistics,* 107(6), pp. 1501–1517.

Bell, C. & Wegner, N., 2025. Feminist Foreign Policy as a Case of Governance Feminism: Neoliberalism, Militarism, and Women as "Agents of Change". *International Studies Review,* 27(1), viaf007.

Bessell, S., 2016. Good Governance, Gender Equality and Women's Political Representation: Ideas as Points of Disjuncture. In: K. Rubenstein & K. G. Young, eds. *The Public Law of Gender: From the Local to the Global.* Cambridge: Cambridge University Press, pp. 273–295.

Bratton, K. A. & Ray, L. P., 2002. Descriptive Representation, Policy Outcomes, and Municipal Day-Care Coverage in Norway. *American Journal of Political Science,* 46(2), pp. 428–437.

Bryant, S., Phillips, J. & Artiz Prillaman, S., 2025. *Gender Equality and Prospects for Good Governance*. London: International Growth Centre.

Buckingham, S., 2010. Call in the women. *Nature,* 468, p. 502.

Celis, K. & Childs, S., 2020. *Feminist Democratic Representation.* Oxford: Oxford University Press.

Celis, K. & Childs, S., 2023. From Women's Presence to Feminist Representation: Second-Generation Design for Women's Group Representation. *European Journal of Politics and Gender,* 6(3), pp. 359–376.

Celis, K., Erzeel, S., Mügge, L. & Damstra, A., 2014. Quotas and Intersectionality: Ethnicity and Gender in Candidate Selection. *International Political Science Review,* 35(1), pp. 41–54.

Chattopadhyay, R. & Duflo, E., 2004. Women as Policy Makers: Evidence from a Randomized Policy Experiment in India. *Econometrica,* 72(5), pp. 1409–1443.

Christoffersen, A., 2021. The Politics of Intersectional Practice: Competing Concepts of Intersectionality. *Policy & Politics,* 49(4), pp. 573–593.

Christoffersen, A. & Siow, O., 2025. There Is No Such Thing as 'Women's Representation': Intersectionality and Second-Generation Gender and Politics Scholarship. *European Journal of Politics and Gender,* 18(3), pp. 670–691.

Collier Hillstrom, L., 2019. *The #MeToo Movement.* Santa Barbara, CA: Bloomsbury Publishing.

Corner, L., 2005. *Gender-Sensitive and Pro-Poor Indicators of Good Governance.* New York: UNDP.

Cornwall, A. & Goetz, A. M., 2005. Democratizing Democracy: Feminist Perspectives. *Democratization,* 12(5), pp. 783–800.

Cornwall, A. & Molyneux, M., 2006. The Politics of Rights: Dilemmas for Feminist Praxis: An Introduction. *Third World Quarterly,* 27(7), pp. 1175–1191.

Cornwall, A. & Nyamu-Musembi, C., 2005. Why Rights, Why Now? Reflections on the Rise of Rights in International Development Discourse. *IDS Bulletin,* 36(1), pp. 9–18.

Crenshaw, K., 1989. Demarginalizing the Intersection of Race and Sex: A Black Feminist Critique of Antidiscrimination Doctrine, Feminist Theory and Antiracist Politics. *University of Chicago Legal Forum,* 1989(1), pp. 139–167.

Dankelman, I. & Fansen, W., 2010. Gender, Environment and Climate Change: Understanding the Linkages. In: I. Dankelman, ed. *Gender and Climate Change: An Introduction.* London: Routledge, pp. 21–54.

Dodd, K. & Schroeder, E., 2024. Policy Blind Spots: A Black Feminist Critique of Sex Trafficking Policy (Inspired by Sara Carrigan Wooten). *Child Welfare,* 102(2 Special Issue: Human Trafficking and Child Welfare), pp. 1–22.

Dugarova, E., 2018. *Gender Equality as an Accelerator for Achieving the Sustainable Development Goals.* New York: United Nations Development Programme.

ECOSOC, 1997. *Gender Mainstreaming: Extract from Report of the Economic and Social Council Report 1997.* New York: UN.

ECOSOC, 2021. *Women's Full and Effective Participation and Decision-Making in Public Life, as Well as the Elimination of Violence, for Achieving Gender Equality and the Empowerment of All Women and Girls.* New York: United Nations Economic and Social Council.

Engle Merry, S., 2011. Measuring the World: Indicators, Human Rights, and Global Governance. *Current Anthropology,* 52(Supplement 3), pp. S83–S95.

Esquivel, V. & Sweetman, C., 2016. Gender and the Sustainable Development Goals. *Gender & Development,* 24(1), pp. 1–8.

FAO, 2011. *The State of Food and Agriculture 2010–2011. Women in Agriculture: Closing the Gender Gap for Development.* Rome: Food and Agriculture Organization of the United Nations.

Goetz, A. M., 2007. Political Cleaners: Women as the New Anti-Corruption Force?. *Development and Change,* 38(1), pp. 87–105.

Halley, J., Kotiswaran, P., Rebouché, R. & Shamir, H., 2018. *Governance Feminism: An Introduction.* Minneapolis: University of Minnesota Press.

Hancock, A.-M., 2007. When Multiplication Doesn't Equal Quick Addition: Examining Intersectionality as a Research Paradigm. *Perspectives on Politics,* 5(1), pp. 63–79.

Harvey, R., 2020. Twitter Reactions to the UN's #HeForShe Campaign for Gender Equality: A Corpus-based Discourse Analysis. *Journal of Corpora and Discourse Studies,* 3, pp. 31–50.

Huyer, S., Acosta, M., Gumucio, T. & Ilham, J. I. J., 2020. Can We Turn the Tide? Confronting Gender Inequality in Climate Policy. *Gender & Development,* 28(3), pp. 571–591.

Inglehart, R. & Norris, P., 2003. *Rising Tide: Gender Equality and Cultural Change around the World.* Cambridge: Cambridge University Press.

International Labour Organization, 2024. *ILOSTAT Database.* [Online] Available at: https://ilostat.ilo.org/data/

Inter-Parliamentary Union, 2025. *Parline Database.* [Online] Available at: https://data.ipu.org/

Inter-Parliamentary Union & UN Women, 2025. *Women in Politics: 2025.*

Kabeer, N., 2015. Tracking the Gender Politics of the Millennium Development Goals: Struggles for Interpretive Power in the International Development Agenda. *Third World Quarterly,* 36(2), pp. 377–395.

King, E. M. & Mason, A. D., 2001. *Engendering Development through Gender Equality in Rights, Resources, and Voice.* Washington, DC: World Bank.

Leach, M., Mehta, L. & Prabhakaran, P., 2016. Sustainable Development: A Gendered Pathways Approach. In: M. Leach, ed. *Gender Equality and Sustainable Development.* Abingdon: Routledge, pp. 1–33.

Leal Filho, W. et al., 2023. Promoting Gender Equality across the Sustainable Development Goals. *Environment, Development and Sustainability,* 25, pp. 14177–14198.

Leftwich, A., 1993. Governance, Democracy and Development in the Third World. *Third World Quarterly,* 14(3), pp. 605–624.

McBride, D. E. & Mazur, A. G., 2013. Women's Policy Agencies and State Feminism. In: G. Waylen, K. Celis, J. Kantola & S. L. Weldon, eds. *The Oxford Handbook of Gender and Politics.* Oxford: Oxford University Press, pp. 654–678.

Mohanty, C. T., 2003. *Feminism without Borders: Decolonizing Theory, Practicing Solidarity*. Durham: Duke University Press.
Nussbaum, M., 2003. Gender and Governance: An Introduction. In: M. Nussbaum et al. eds. *Essays on Gender and Governance*. New Delhi: Human Development Resource Centre, UNDP, pp. 1–19.
OECD, 2019. *Fast Forward to Gender Equality: Mainstreaming, Implementation and Leadership*. Paris: OECD Publishing.
OECD, 2021. *Gender and the Environment: Building Evidence and Policies to Achieve the SDGs*. Paris: OECD Publishing.
OECD, 2025. *Gender Equality in a Changing World: Taking Stock and Moving Forward*. Gender Equality at Work. Paris: OECD Publishing.
Perry, K. & Burns, C., 2023. Remembering the "Human" in Human Trafficking: An Analysis of Female Leadership and Anti-Trafficking Policy Choices. *International Studies Perspectives*, 24(2), pp. 148–168.
Phillips, A., 1998. *The Politics of Presence*. Oxford: Oxford University Press.
Squires, J., 2007. *The New Politics of Gender Equality*. Basingstoke: Palgrave Macmillan.
Stevens, C., 2010. *Are Women the Key to Sustainable Development?*, Boston, MA: Boston University, The Frederick S. Pardee Center for the Study of the Longer-Range Future.
Tarr-Whelan, L., 2010. The Impact of the Beijing Platform for Action1995 to 2010. *Human Rights*, 37(3), pp. 2–23.
Thomas, M. A., 2010. What Do the Worldwide Governance Indicators Measure?. *The European Journal of Development Research*, 22, pp. 31–54.
Tosi, N., 2025. Gender-Based Quotas, Parity, and Elections. *Policy Perspectives*, 32, pp. 1–28.
UNDP, 2025. *Gender Inequality Index*. The Human Development Report. [Online] Available at: https://hdr.undp.org/data-center/
United Nations Women, 2025. *Progress on the Sustainable Development Goals: The Gender Snapshot 2025*.
United Nations, 1995. *The Beijing Declaration and the platform of action. Fourth World Conference on Women*, Beijing, China. UN New York, Department of Public Information.
UNRISD, 2005. Gender and "good governance". In: *Gender Equality: Striving for Justice in an Unequal World*. Geneva: United Nations Research Institute for Social Development (UNRISD), pp. 181–192.
Verloo, M. & Lombardo, E., 2007. Contested Gender Equality and Policy Variety in Europe: Introducing a Critical Frame Analysis Approach. In: M. Verloo, ed. *Multiple Meanings of Gender Equality: A Critical Frame Analysis of Gender Policies in Europe*. Budapest: Central European University, pp. 21–50.
Walby, S., 2005. Introduction: Comparative Gender Mainstreaming in a Global Era. *International Feminist Journal of Politics*, 7(4: Comparative Gender Mainstreaming), pp. 453–470.
Waylen, G., 2008. Gendering Governance. In: G. Goertz & A. G. Mazur, eds. *Politics, Gender, and Concepts*. Cambridge: Cambridge University Press, pp. 114–135.
World Bank, 2024. *Worldwide Governance Indicators*. [Online] Available at: https://www.govindicators.org/
Zinsser, J. P., 2002. From Mexico to Copenhagen to Nairobi: The United Nations Decade for Women, 1975–1985. *Journal of World History*, 13(1), pp. 139–168.

5

DECOLONIAL GOVERNANCE AS GOOD GOVERNANCE

Vanessa Burns

Introduction

What does it mean to decolonise governance, and why would this necessarily represent good governance? Here I explore what decolonisation means for good environmental governance, with an emphasis on postcolonial regions in the Global South. I define environmental governance broadly, as the set of processes, institutions, rules, and practices that guide human interactions with the environment. I position decolonisation as a project of institutional reform that works against colonial logics in international environmental law and governance frameworks (Carter, 2018; Burns, 2019, 2022; Sultana, 2022, 2025). I argue that there is a moral, prudential, and pragmatic imperative to pursue this end: moral, because neocolonialisms further entrench uneven power relations; prudential, because it is in the institutions' best interests to decolonise; and pragmatic, because decolonial governance is better adapted for regional and domestic policy implementation where regions have alternative (non-Western) ways of governing land and sea.

My use of the term decolonisation differs from the formal processes by which former colonies gain political and governmental independence from a colonial power. Instead, the ongoing project seeks to dismantle the deep, enduring structures of coloniality: a matrix of power, knowledge, and practice that perpetuates racial, social, and economic hierarchies. This scholarly and activist movement, largely concentrated in the Global South across Latin America, Africa, and parts of Asia, finds its intellectual roots in the work of key thinkers like Aníbal Quijano (Quijano et al., 2024), Walter Mignolo (2011; see also, Mignolo and Walsh, 2018), and Sabelo J. Ndlovu-Gatsheni (2015, 2016, 2021). Key concepts such as the 'coloniality of power' and the 'coloniality

DOI: 10.4324/9781003467694-6

of knowledge' describe the continuity of colonial power hierarchies and critique the Eurocentric assumptions that continue to underpin global governance models. The movement's primary focus is not simply political liberation, but on a radical epistemic and ontological shift towards what Arturo Escobar (2011, 2018) calls a pluriversal world.

From a social change perspective, we can situate decolonial governance relative to either radical or reformist approaches; either reliant on deep structural change (see Fanon, 1967; Collins, 2024) or, as poststructuralists argue, reliant on reforming existing institutions (see Carter, 2018). There are also normative questions concerning what decolonial governance should achieve. Is it a material process concerned with ongoing settler colonialism and repatriation of native land (Tuck and Wayne Yang, 2012; Curley and Gupta, 2022)? Is it instead a metaphor for a broader social justice movement (Garba and Sorentino, 2020)? A spectrum of theoretical positions in the decolonial literature offers points of departure. These positions can be summarised in two ways: on one end, a backward-looking position that sees decolonisation as the undoing of harms inflicted on colonised lands, peoples, and their descendants; on the other, a forward-looking position that considers decolonisation as a process of addressing contextual injustice through grassroots epistemic change.

The argument for global governance of global concerns is compelling in a time of catastrophic climate change. Yet this is currently pitted against aims to decolonise environmental governance, because global governance regimes (such as the environmental programmes run by the UN) are deeply Westphalian in their conception. When Global South nations are asked to comply with global governance regimes, by conforming to European-style knowledge systems in the conception of local and regional governance frameworks, these global regimes become neocolonial forms of governance. This chapter is concerned with good environmental governance which as an ideal brings together several empirical and normative aims. I take a rights-based interpretation of good governance as my starting point and argue that neocolonial forms of governance fail to uphold the cultural rights of Indigenous People, by undermining alternative knowledge systems. If we take a reformist approach to decolonising environmental governance, and I argue here that we should, then the work of institutional change falls to any scholars, legislators, policy makers, or community leaders contributing to institutional practices, either local or supralocal.

This chapter draws from case studies carried out in the equatorial Pacific, where decolonial efforts often focus on reclaiming Indigenous knowledge and confronting specific neocolonial challenges related to climate change, resource extraction, and sovereignty, as articulated by scholars such as Linda Tuhiwai Smith (Smith, 2012; Smith et al., 2018) and Konai Helu Thaman (2003, 2019).

What follows is a consideration of theoretical approaches to decolonial governance, and an argument for a related supracolonial approach. I then discuss an empirical case that illustrates a central problem facing efforts to decolonise environmental governance. After this, I share two examples of decolonial (or supracolonial) environmental governance; a discussion of state-level iTaukei (Indigenous) governance in Fiji, and a community-level example from Timor-Leste. Lastly, I reflect on the problem of anthropocentrism in environmental governance and set out my argument for why this supports the case to decolonise governance concepts and practices in regions where strong ontological alternatives exist.

Theoretical approaches to decolonial environmental governance

For as long as postcolonial nations in the Global South are asked to comply with international law and governance frameworks, or remain internally colonised by European-style nation-states, there remains an urgent call for these frameworks to be decolonised. The violent legacies of colonialism in these regions, and the ways in which colonial logics implicit in European frameworks can *reproduce* epistemic violence, make this a moral imperative. Yet there are also pragmatic reasons; colonial logics obstruct good governance in the Global South, lead to maladapted governance regimes, and problems with participation, implementation, and compliance. For example, colonial logics implicit in international treaties like the United Nations Convention on the Law of the Sea (UNCLOS) can fundamentally obstruct the development of reformist policy approaches. This is because regional, national, and local marine governance instruments are forced to adopt European-style governance logics to comply with UNCLOS, despite them being maladapted to Indigenous, local governance regimes (Burns, 2019, 2022). Similarly, Yolanda Ariadne Collins (2024) argues that the United Nations Reducing Emissions from Deforestation and Degradation (REDD+) initiative is fundamentally ineffective in the Amazonian Guiana Shield, because it fails to account for colonial histories that have positioned forests as sites of refuge from and resistance to colonial violence for Indigenous and Afro-descendant communities.[1]

The challenge, as Paul Carter (2018) puts it, lies in recognising and valuing alternative Indigenous knowledge systems. For Carter, this means avoiding reducing Indigenous knowledge to mere local perspectives (as can happen when Indigenous knowledge is studied *ex situ*), and instead seeing these knowledge systems as situated philosophical approaches. Drawing from research in the Indonesian archipelago, Carter (2018) argues that decolonising governance requires a remapping of human relations. He points to a gap in decolonial environmental scholarship of oceanic regions to suggest an 'archipelagic approach' that promotes a relational theory of social organisation that centres Indigenous knowledge (and thus, I would argue, decentres humans in theories of social organisation).

The examples set out below are reflections on my own work, undertaken in the Indonesian archipelago and Melanesia over the last 15 years to investigate decolonial approaches to climate adaptation governance. By revisiting this work, I hope to illustrate problems in domestic and regional environmental governance as symptomatic of problems elsewhere, as well as offer empirical examples of supracolonial approaches to environmental governance.

A supracolonial project

European-style systems of law and governance embody and reproduce European Enlightenment era ideas, as would be expected of any historically founded and long-practised knowledge system coming out of a particular social, economic, and political tradition. However, the colonial project, globalisation, and global capitalism have resulted in a world in which ideas of European origin now dominate governance not just in Europe but across both hemispheres. These ideas are not static and explicit; they are dynamic and implicit. They are so embedded in institutional practice that their persistent reproduction, through ontological principles (i.e. implicit rules derived from a particular set of logics) is difficult to isolate and observe (Burns, 2022). This could mean that problematic principles implicit in dominant governance regimes are beyond reform. From a structuralist position, the institutional reproduction of colonial logics instead requires radical transformation. Fanon (1967 and elsewhere) argues that decolonisation can only be achieved if existing institutions, often imitations of the coloniser's, are replaced with new, Indigenous ones. Even this might not be enough to meet the fundamental aim of the decolonial project – to decolonise.

This points to a fundamental flaw in the decolonial concept itself: we cannot undo colonialism. We cannot, in effect, decolonise. This is true in two senses: most obviously, we cannot put colonialisms of the past back in the box. Second, colonialism is not a static set of ideas and practices situated in the past; these ideas and practices persist and reproduce in contemporary institutions. This happens both through human agency *and independent of it*, via rules, routines, discourses, and material structures. From this position, a *non*colonial world or people can never exist. When Fanon asks us to 'look at how the shadow of Europe's monuments spreads and multiplies' (1967, 111) he is asking us to think about the boundless reproduction and reappropriation of expansionist, extractive, dualistic principles and practices of colonialism's legacy.

The word decolonial literally means a process by which the colonial is separated *from*, made *apart from*, the noncolonial.[2] If decolonisation is not a metaphor (see Tuck and Wayne Yang, 2012; Garba and Sorentino, 2020) then what we might realistically strive for is a supracolonial project. I use the prefix supra to mean 'situated over… [or] upon another' (OED, 2025) to describe the act of living over or upon the remnants of colonialism, which is (both literally and metaphorically) so often the case in postcolonial nations. This is in essence a reformist position arising from the pragmatics of needing existing global institutions to facilitate alternatives in a time of environmental crises, but one that requires fundamental transformation of these institutions. Secondly, it rests on the argument that any radical decolonial alternative is likely to reproduce colonial logics, because of the pervasiveness of European thought and the difficulty of rooting out ontologies that 'have become a naturalised mode of thinking' (Peters, 2020, 5).

A supracolonial project works against the colonial not by rooting it out (as the structuralists say), and not by simply doing things differently (as the post-structuralists say), but by observing, remembering, and making sense of these 'shadows' of colonialism. I take the term supracolonial from Ghassan Hage's supra-counter-colonialism. Hage, a Lebanese Australian, introduces this term in a piece written for *The Australian* newspaper (Hage, 2008) in which he recalls passing through Bathurst, a rural frontier town in Australia, and stopping to visit an overgrown garden that once belonged to his grandfather:

> ...there, amid the chaos, I could discern three unmistakable forms: a fig tree, an olive tree and a pomegranate tree, the holy Mediterranean trinity, or one of them at least... The Lebanese trees did not make me feel Australian and Lebanese...Nor did they make me feel torn between my Lebaneseness and my Australianness. They simply made me feel...more Australian...For many people, a greater sense of rootedness does not mean a sense of being locked in the ground, unable to move. On the contrary, roots often are paradoxically experienced like an extra pair of wings. And this was exactly how I experienced my trees. I felt them propelling me...they are not roots that keep you grounded, they are roots that stay with you as you move... this is not an anti-colonial belonging, which pits the belonging of the colonised against that of the coloniser while conserving colonialism's either-or logic. Nor is it a post-colonialism, which prematurely sees colonial culture as something superseded. If anything, it is a supra-counter-colonialism: it counters colonial culture from a space outside of and beyond it.

This 'supracolonial' process of remembering the past, making sense of it, and using it to propel you forward describes so much of what I see when working with communities in postcolonial small island developing states. I am especially thinking of communities in Timor-Leste, who live amongst the literal ruins of the Indonesian occupation, whose involuntary memories are often driven by PTSD because of intergenerational colonial violence. Here, the crumbling walls of internment camps are now maintained as seawalls to protect villages against the rising tides, and NGO-installed water tanks are integrated into an animistic custodianship over the groundwater. In the remaining sections I explore why neocolonialism is a problem in environmental governance and consider two alternatives as supracolonial approaches.

Whitewashing maladaptation as Indigenous 'lack of capacity'

Environmental governance in the Global South is now inextricably linked to development. Yet, while the Sustainable Development Goals (SDGs) represent major reforms towards good environmental governance, this most recent model of international development still promotes a liberal growth economy and relies on concepts like 'natural capital' and 'ecosystem services' that position natural resources as passive categories to be (sustainably) exploited. This is an idea fundamentally at odds with most Indigenous models of environmental management which instead position things in nature as part of the social fabric. This relational approach instead decentres the human, and holds individuals, communities, and nations morally accountable. This necessitates a non-market substantive economic model that relies on reciprocity, redistribution, and, as economic historian Karl Polanyi (1944/2024) notes, social and ecological embeddedness. International development governance is, therefore, maladapted to Indigenous models in at least two ways: it promotes a global market economy that is disembedded from ecological concerns; and it relies on categorisations of nature that are disembedded from the moral/social fabric. From a decolonial perspective, both the liberal growth economy and dominant categorisations of nature have their origins in the European Enlightenment. Both are incongruent to the local Indigenous practices of many postcolonial nations, and imposing these ideas through development paradigms is an act of neocolonialism (Murrey and Daley, 2023).

Yet, these ideas and practices are so embedded in development practice, and the mechanisms of international governance more generally, that practitioners and institutions are caught in a cycle of reproduction that evades attention and defies reformist efforts (Burns, 2022). This reproduction of European ideas in postcolonial settings is what decolonial theorists refer to as epistemic violence (Spivak, 1988). What these styles of governance require of Indigenous Peoples is not simply change, or reform, or the temporary adoption of a different set of ideas to achieve a development end. In practice, they actively undermine, ridicule, and erase Indigenous knowledge and practice. They dispossess communities and nations of intergenerational knowledge, and they make this the price of participation in development.

What does this look like in practice? Although there will be many different versions and experiences that offer answers to this question, they are united in at least one way: that neocolonialism looks and feels like business as usual. Colonial logics are implicit and normalised in the governance concepts, instruments, frameworks, and institutions that we use for development, and environmental governance.

In 2013, I attended the 3rd Regional Exchange (REX) on Climate Change Adaptation in Dili, Timor-Leste, held by the Coral Triangle Initiative on Coral Reefs Food and Fisheries (CTI-CFF). This is a non-binding multilateral agreement between the six member nations: Malaysia, Indonesia, the Philippines, Timor-Leste, Papua New Guinea, and the Solomon Islands. Together, the Coral Triangle nations are home to roughly 120 million people, many of whom are Indigenous Peoples with subsistence-based fishing livelihoods. The REX was led by the United States Agency for International Development (USAID) and involved a week of formal meetings and facilitated group work. I was there to conduct ethnographic research of the REX itself, as part of broader research on adaptation to climate change in Timor-Leste. On the REX agenda was the launch of a new adaptation policy tool, the *Climate Change Adaptation for Coral Triangle Communities: A Guide for Vulnerability Assessment and Local Early Action Planning (LEAP Guide)*. This was designed to feed local data into the implementation of national adaptation plans, such as Timor-Leste's first National Adaptation Programme of Action (NAPA). However, multilateral policy exchanges are never simply sites where policy is announced or developed. They are political arenas in which hierarchical relations are rehearsed and institutional epistemes are enacted. Encounters are mediated by frameworks, templates, and instruments that travel with aid programmes and structure what counts as legitimate knowledge (Lewis and Mosse, 2006). Classic critiques, such as those made by Escobar (1995), show how these instruments translate complex local realities into administrable categories, shaping what delegates discuss and prioritise at meetings. Networks and hierarchies are also established that extend beyond single events, such as those documented by Pietri et al. (2015) in a study of how CTI-CFF regional exchanges produce emergent networks that shape regional governance trajectories.

What emerged during the REX was a logic of differentiation that positioned Papua New Guinea, Timor-Leste, and the Solomon Islands as 'lacking capacity.' In the immediate, this referred to the capacity for delegations to meaningfully engage in CTI-CFF led policy implementation activities at the REX. However, the general discourse also implied a longer term 'lack of capacity' to meaningfully engage in regional governance aims. This framing – repeated across plenary discussions and informal exchanges – reproduced colonial hierarchies of competence and legitimacy within the regional initiative. On the first day of the workshop, for instance, I was assigned by the organisers to assist the Timorese delegation in completing the group exercises, a gesture that both reflected and reinforced the assumption of incapacity. During Session 4 on 'Lessons from Community Adaptation,' an environmental officer from Timor-Leste resisted this narrative, drawing attention instead to the maladaptation of the LEAP instrument itself.[3] He noted that, although community consultations had been undertaken in the tool's design, they were insufficient and superficial, resulting in a rejection of both the training activities and subsequent adaptation initiatives. His intervention underscored that what was framed as a deficit of local capacity was, in practice, a failure of institutional reflexivity – a refusal to recognise the epistemic and practical limits of the tool itself. I trace my scholarly interest in decolonial governance back to this roundtable discussion. How could policy be developed in consultation with local communities, yet produce a toolkit so maladapted that it was effectively unusable? Perhaps more importantly, why was a lack of practitioner participation in the REX, and of community participation in the toolkit being framed as 'lack of *capacity*'?

The officer's critique was absent from the 'best practices' identified in the REX Summary Report. What was included, however, hints at the same underlying problem:

> …Indonesia and Solomon Islands related that participants found the LEAP training 'too long and complicated'…Participants suggested that (a) the LEAP should be adapted to local needs; (b) the activity should not last more than three days; and (c) the information should be presented in a way that is easy for fishers to understand (i.e., using the local language, not science language). Indonesia has developed new guidelines based on the *LEAP Guide* "but modified for our purpose and not too complicated" …PNG [Papua New Guinea] has also initiated their own local community-based CCA [climate change adaptation] planning, which the government coordinates through their NGO partners – the plans are each written in the community's own language. Community-based adaptation appears to be the preferred approach in most countries. Timor-Leste noted that "communities have done their own climate adaptation, even if they don't understand what climate change is" … Solomon Islands in particular described community engagement as 'a big challenge.'[4]

Two points stand out. First, that those member nations mentioned approach adaptation through community-based and decentralised systems of environmental governance, reflecting the customary structures that predate colonial rule. In Timor-Leste, for instance, governance is organised around precolonial kingdoms, subsistence livelihoods, and animistic custodial practices. Second, that the regional policy instruments produced under the CTI-CFF framework were profoundly maladapted to these local systems. Indonesia and Papua New Guinea developed alternative state-level instruments in response, while Timor-Leste and the Solomon Islands communities developed their own informal or customary approaches in parallel.

What is at play here is a *whitewashing of maladaptation* as 'lack of capacity.' Within policy discourse, the term *capacity* is often used to refer to institutional, human, or financial resources. But when directed at a people or entire nations, it acquires a distinctly racialised dimension – implying cultural or cognitive deficiency rather than structural incompatibility. I have come to understand it as a deeply racist term, directed at those cultures in the least developed nations (such as Timor-Leste and Solomon Islands) whose systems of governance least resemble European models. The language in the REX report reflects this legacy. Describing training as 'too long and complicated' infantilises participants, implying short attention spans or limited understanding. The belated recognition that materials should be translated into local languages betrays the superficiality of consultation. At the time of research, most elder community leaders in Timor-Leste, for example, were oral custodians who neither read nor wrote in the official or working languages of development agencies (Portuguese, Tetum, English, or Indonesian). Many spoke only one of Timor-Leste's 30 traditional languages. This linguistic and cultural distance rendered the policy tool inaccessible: not because communities lacked capacity, but because the instrument itself was conceptually and practically misaligned. In Timor-Leste, the toolkit was not merely impractical but actively rejected, as it conflicted with customary practices of environmental management. Here, 'lack of capacity' becomes a rhetorical device used to obscure institutional failure. Continuing to impose European-style governance models and requiring 'participation' in their implementation demands not only that communities learn new environmental concepts, but that they internalise them. This not only competes with traditional management systems but also works towards their erasure. This is a profoundly colonial dynamic, in which the price of aid is the adoption of the coloniser's epistemologies of governance.

At the heart of this problem lies the Enlightenment-era dualism between humans and nature: a worldview that positions the environment as a passive, external resource. Governance tools built upon this ontology will always be maladapted to animistic custodial systems, where human-environment relations are relational, dynamic, and reciprocal. The issue, therefore, is not one of 'capacity' but of conceptual dissonance. Participation in such frameworks requires Indigenous users to perform a radical Westernisation of their own governance systems to qualify as legitimate participants in regional adaptation regimes. The ethnographer and theorist Stephen Muecke (Muecke and Roe, 2021) describes this process through the concept of the *knowledge valve*. During the colonial period, Europeans extracted local knowledge and translated it into their own epistemic frameworks, transforming knowledge 'grown' from Country (Muecke and Roe, 2021, 23) into knowledge that 'fits' European logics.[5] The valve is one-directional: knowledge passes from Indigenous contexts into European systems, where it is reclassified and decontextualised, while European systems themselves remain unexamined. Let us imagine, for instance, if the process worked in reverse: if European communities were required to co-produce an environmental governance tool grounded in animistic custodial logics. Their private property laws, fixed boundaries, and linear territorialities would prove maladapted to a system that recognises no private ownership, no straight borders, and regards all things as relational and alive. Implementing such a tool would demand a fundamental reorganisation of their environmental governance – precisely what is asked of Indigenous and postcolonial communities under current sustainable development models. Under this logic, nations are deemed to have 'capacity' in proportion to how well their governance systems mimic European institutional forms. Those that resist such mimicry are racialised as deficient. In this sense, the discourse of 'capacity' does not measure readiness or skill; it measures proximity to colonial epistemology.

Decolonial governance

In this section, I offer two examples of decolonial alternatives. First, I draw from community research in Timor-Leste to show how locally planned approaches to adaptation governance demonstrate a supracolonial approach to environmental governance in the Coral Triangle. Secondly, I draw from iTaukei (Indigenous) systems of governance in Fiji to consider how these institutions sit both within and against the nation-state. What these examples offer is a reflection on the subversion of power hierarchies where de/colonial ontologies coexist in postcolonial settings. What both examples have in common is strong, decentralised, Indigenous governance systems, and customary land tenure, significantly strengthening the structural basis for alternative systems of governance. I also consider the Fijian concept of *vanua* and the Timorese concept of *lulik* as central to defining why these are alternative, and decolonial, systems of environmental governance.

Animistic supracoloniality in Timor-Leste

As noted above, the discursive use of the term 'lack of capacity' offers insight into the reproduction and use of colonial hierarchies in Global South environmental governance. The LEAP tool's failure in Timor-Leste was not a case of local deficiency, but of conceptual misalignment. To look beyond this failure, my subsequent fieldwork in Timor-Leste and elsewhere in the Coral Triangle has turned to the adaptive practices already embedded within Indigenous societies: practices that were not only surviving but generating coherent systems of environmental governance. What I encountered was not resistance to regional governance, but a different order of governance altogether. I describe this here as supracolonial governance – a mode of local governance that exceeds both colonial and postcolonial systems. These practices persist not in opposition to the remnants of colonialism but in conversation with them, drawing upon Indigenous ethical and spatial logics that seek to reconfigure rather than reject.

The example I offer here is the custodianship of sacred springs (*uma lulik*) by (primarily) women elders in Timor-Leste. In the animistic cosmologies of the Timorese, these springs are regarded as part of the ancestral landscape with agency and intention of their own – entities that sustain life when properly cared for and withdraw it when neglected. Their maintenance involves both ritual and pragmatic practices. Each act embodies the principle that water is not a resource but a lively part of the social and moral fabric. As climate variability intensifies, these acts of care take on new ecological significance. When springs dry or shift, women's walking practices between village and source lengthen, tracing not just a hydrological change but a social geography of adaptation. In this sense, women's movements both delineate systems of groundwater and inscribe the contours of the drought (Burns, 2019).

This form of adaptation is not reactive but anticipatory, grounded in relational observation accumulated over generations. For instance, when a spring's taste changes from fresh to slightly saline, women interpret it as a message: an index of imbalance that requires ceremonial repair and ecological adjustment. The community may respond by replanting bamboo along stream edges or relocating agricultural plots to more shaded slopes. These actions arise from the same epistemic framework that understands water, soil, and people as interdependent kin. Within this system, governance is not a structure but a practice of continuous negotiation among human and non-human actors. The authority of the women who maintain this balance is both social and ecological: they allocate access to water, time rituals around rainfall, and ensure that agricultural and domestic uses do not disrupt the moral equilibrium of the landscape. Their decisions, though locally grounded, carry regulatory force; they shape where families plant, build, and move.

To view these practices through the lens of supracoloniality is to recognise how they convert the residues of colonial disruption into resources for survival. As Hage (2017) suggests, decolonisation involves not only resisting colonial histories but also using them. In Timor-Leste, this process unfolds through an animistic ethics of attention. *Observing* the sedimented legacies of colonial rule in degraded landscapes, *remembering* ancestral practices through story and ritual, *making sense* of contemporary environmental shifts within an ontology that attributes agency to the more-than-human world, and *using* these understandings to adapt. These are the steps through which women translate colonial aftermaths into living systems of care. Each act of maintenance and movement reclaims the colonial terrain, repurposing it for Indigenous life. In doing so, these women develop an autochthonous method – a supracolonial method grounded in continuity and creative transformation.

From the perspective of external frameworks like the LEAP tool, these practices remain illegible because they do not conform to the representational norms of governance. They are not quantified, mapped, or written into reports; they are enacted through dynamic practice that moves over space and time. When the LEAP process judged such communities to be lacking capacity, it was, in effect, mistaking opacity for absence. But opacity can be a form of resistance. By refusing to translate embodied knowledge into the bureaucratic categories of *resilience* or *vulnerability*, these women preserve epistemic autonomy that development frameworks can neither comprehend nor co-opt. Their walking, in this light, is a supracolonial practice of governance – one that redefines capacity not as compliance with external metrics but as the capacity to sustain relationships across human and ecological boundaries. These insights speak directly to the dynamics observed at the REX. What donor representatives perceived as disengagement or incapacity was, in fact, a refusal to participate in a system that sought to abstract and quantify relations that are, by their nature, embodied and reciprocal. To call this a lack of capacity is to invert the truth: what was lacking was the capacity of external governance frameworks to perceive the relational and ethical dimensions of Indigenous environmental management.

This lesson resonates beyond Timor-Leste. As the next section will show, the institutionalisation of iTaukei governance in Fiji also embodies the entanglement of colonialities and Indigenous resurgence. Like women's water custodianship in Timor, iTaukei governance navigates the tension between bureaucratic form and relational substance – between the universalising logic of the nation-state and the situated ethics of *vanua*. Together, these cases illuminate how supracolonial governance emerges not from institutional innovation but from the enduring capacity of Indigenous systems to inhabit, repurpose, and exceed the architectures of empire.

ITaukei governance and the Fijian state

Fiji's governance system operates through a hybrid structure that overlays customary authority and communal land tenure with modern statutory institutions. ITaukei governance is, therefore, not a discrete subsystem of the state; it is a parallel and constitutive mode of governance that shapes how the state manages land, resources, and community life.

Institutions such as the Ministry of iTaukei Affairs (MTA), the iTaukei Land Trust Board (TLTB), and the iTaukei Lands and Fisheries Commission (TLFC) embody this hybrid system. Established under the iTaukei *Affairs Act* (Cap. 120), iTaukei *Lands Act* (Cap. 133), and *iTaukei Land Trust Act* (Cap. 134), they administer and translate customary principles within statutory law. The TLTB manages leases of iTaukei land on behalf of landowning groups, the TLFC adjudicates disputes and maintains registers of land and fishing grounds, and the MTA oversees broader policy concerning iTaukei welfare and traditional governance.

Two interrelated features of iTaukei governance are communal land tenure and the concept of *vanua* – a cosmology that sees land, sea, people, and ancestors within a single, living system. For ITaukei people, customary tenure of land is a birthright administered by the iTaukei Land Trust Board (TLTB), with approximately 80% of Fiji's land held by the landowning units around which iTaukei society is organised: *mataqali* (extended family), *yavusa* (clan), and *tokatoka* (tribe). These collective forms of ownership are not merely legal arrangements; they express an ontology of relational custodianship that contrasts sharply with colonial and capitalist notions of private property, resource extraction, and alienable rights. These iTaukei institutions are integrated into the state in several ways. The Constitution of Fiji recognises customary iTaukei law in relation to land ownership (e.g. Chapter 2 of the Bill of Rights, especially Sections 28 and 29). The relevant statutes give legal effect to customary ownership and practices (e.g. acknowledging that iTaukei land is inalienable) while placing them within a framework of state oversight (e.g. roles of TLFC, TLTB) (World Bank, 2017; Parliament of Fiji, 2017). *Vanua*, in turn, gives spiritual and ecological depth to these land relations, through recognition of iTaukei culture, customs, and traditions in the Constitution's Preamble, and as a working concept in Ministerial practice.[6]

However, there are tensions and trade-offs. The incorporation of iTaukei institutions within the Fijian state has often been framed as a recognition of Indigenous rights (Nayacakalou, 1975). Yet decolonial theory prompts a different reading. State oversight does sometimes limit pure customary autonomy (e.g. statutory limits on how land may be leased or restrictions on free sale of customary land), while demands of transparency, uniform legislation, accountability, and their integration into national development agendas impose bureaucratic norms that are in tension with customary governance practices (Ward, 1995; Nayacakalou, 1975). This aligns with arguments that postcolonial states frequently continue forms of internal colonisation, whereby Indigenous peoples are subordinated within national structures, their lands incorporated into state frameworks, and their knowledge subsumed under Eurocentric legal and administrative regimes (Quijano, 2000; Smith, 2012). In Fiji, although the 2013 Constitution guarantees communal land tenure and recognises iTaukei identity, it also centralises authority in the state and subjects Indigenous institutions to ministerial oversight.

Yet, rather than seeing these institutions only as administrative arms of the state, we can view them as sites of tension and struggle over epistemic and political sovereignty, where iTaukei concepts and practices of governance work both within and against the modern nation-state. In this light, the iTaukei institutions can be read in two ways: as instruments of incorporation (bringing Indigenous land, sea, and leadership structures under state bureaucratic management) but also as arenas where iTaukei concepts such as *vanua* and *mataqali* persist and challenge the territorialising functions of state sovereignty. Far from denoting territory, *vanua* situates land, sea, people, and ancestors as an interconnected ecology that is relational and pluriversal – rooted in a worldview where land is not a commodity but a living, sacred foundation of community. When TLTB administers leases, or when TLFC demarcates boundaries, they are operating in a space where *vanua* persists as an implicit principle. Even though these agencies function through statutory law, they continually confront claims articulated in *vanua* terms: disputes are not only over parcels of land but over ancestral relations, obligations to community, and sacred responsibilities. From a decolonial perspective, *vanua* resists the commodification and abstraction of land that characterises colonial and capitalist systems. It insists that governance must account for spiritual and communal responsibilities, not just market value or state efficiency. Thus, *vanua* is a form of 'epistemic disobedience' (Mignolo, 2009) that asserts a different communal ontology that works against colonial territorialist logics and the legal frameworks through which these logics are upheld.

The principles of *mataqali, yavusa*, and *tokatoka* also work against liberal property regimes premised on individual ownership and alienability. In contrast, iTaukei tenure is communal, perpetual, and inalienable. The operation of the TLFC also demonstrates how customary law persists within statutory structures. By recording disputes 'according to …native custom and usage' (Native Lands Act, Cap 133, Section 3) the Commission acknowledges a source of legitimacy outside the state. Though mediated by bureaucratic procedure, this recognition is important: it allows Indigenous concepts of justice, leadership, and land use to enter state records. This is not without contradictions. By codifying custom into registers, the Commission risks freezing flexible and context-specific traditions into rigid categories – what anthropologists call the 'museumisation' of custom (Jolly, 1992). Yet, the survival of these registers, and the requirement that state courts consult them, ensures that iTaukei legal principles remain embedded within national governance. This aligns with decolonial calls for legal pluralism as a way of unsettling colonial homogeneity (Merry, 1988; Pahuja, 2011).

Taken together, iTaukei concepts and Fijian institutions illustrate the persistence of internal colonisation operating alongside decolonial alternatives. The integration of iTaukei governance into the Fijian state both exemplifies and complicates Murrey and Daley's (2023) critique of the colonial nation-state model. The institutionalisation of iTaukei structures attests to the endurance of colonial administrative forms. Each of these institutions was conceived within a British framework of indirect rule. This maintained chiefly authority and communal land tenure only insofar as they could be rendered legible to, and managed within, a centralised bureaucracy. In this sense, iTaukei governance remains embedded within the homogenising architecture of the nation-state:

it is exercised through statutory instruments, ministerial oversight, and codified registers that reproduce what Pahuja (2011) describes as the 'universalising logic' of modern legality.

Yet, at the same time, iTaukei governance also gestures towards what Murrey and Daley call *alternative imaginaries*. The recognition of *vanua* (land-community) and *veiwekani* (relationality) within official discourse resists purely territorial or economic conceptions of governance and reasserts Indigenous epistemologies of care and belonging. While these institutions remain constrained by the postcolonial state and its integration into global capitalist structures, their existence continues to unsettle colonial homogeneity by maintaining a plural, legal, and moral order within the nation. Thus, iTaukei governance simultaneously reproduces and subverts the colonial architecture of the state, embodying both the persistence of empire's bureaucratic legacies and the decolonial aspiration to govern otherwise.

The problem of anthropocentrism in environmental governance

The iTaukei concept of *vanua* and the Timorese concept of *uma lulik* provide a necessary counterpoint to the dominant problem in environmental management: anthropocentrism. Anthropocentrism – the material and moral centring of human interests – is not merely a philosophical error; it is the practical ideology that justified and facilitated the colonial project of resource extraction and territorial domination.

Drawing on the argument that a specific ideology (X) leads to anthropocentrism, which in turn causes environmental crisis (Brennan and Norva, 2021), we can posit the following theoretical link:

1 Coloniality (X) leads to anthropocentrism;
2 Anthropocentrism leads to environmentally damaging behaviours;
3 Therefore, coloniality leads to environmental crisis.

The philosophical foundations of colonialism, rooted in European Enlightenment rationalism, simultaneously elevated the rational human subject while reducing colonised peoples and their lands to the status of irrational non-human other. This created the moral license to strip both their intrinsic value and autonomy, paving the way for systematic resource exploitation.

The decolonial governance models examined here directly challenge this legacy. Concepts like *vanua* represent an inherent constitutional ecocentrism that resists the commodification of nature. Similarly, the supracolonial and relational custodianship of water in Timor-Leste actively refuses to translate embodied knowledge into the purely managerial and utilitarian language of adaptation frameworks. By insisting that governance must account for sacred, communal, and non-human responsibilities, these Indigenous systems challenge the territorial and anthropocentric logics inherited from the colonial nation-state.

Conclusion

Ultimately, this chapter has argued that the project of decolonising environmental governance is not merely an academic or political exercise, but a moral, pragmatic, and prudential imperative. We have explored how the enduring 'matrix of coloniality' continues to operate within international environmental law and governance, perpetuating a harmful and maladapted system that is fundamentally at odds with the goals it purports to achieve. By tracing the lineage of this problem from colonial logics to a deeply embedded anthropocentrism, I have demonstrated how global governance frameworks can inadvertently delegitimise and erase effective, place-based Indigenous alternatives. The case of the Coral Triangle Initiative highlights this dynamic, showing how the 'lack of capacity' narrative serves as a rhetorical device to obscure the fundamental inadaptability of Euro-centric policy instruments to the realities of a pluralistic, postcolonial world.

The problem, as we have seen, is not a failure of will on the part of individual practitioners, but a structural and institutional flaw rooted in the very ontological principles that define the Euro-Western worldview. This chapter has offered a critique of this one-way 'knowledge valve' that extracts and assimilates Indigenous knowledge while refusing to allow its own frameworks to be fundamentally transformed. In response to this, I have proposed a 'supracolonial project': a reformist position that acknowledges the pervasive legacies of colonialism while charting a path for living over and upon them. For scholars and policymakers, this means an active and engaged commitment to institutional change, engaging meaningfully with epistemic alternatives, and confronting the colonial shadows that continue to obstruct good environmental governance.

The examples of Indigenous governance discussed here demonstrate a less anthropocentric (and therefore more sustainable) approach to managing resources. In this sense, they are decolonial because they represent alternatives forms of governance, and they are 'good' because they are more effective ways of achieving sustainability. This suggests that where these alternative ontological systems are embedded, the logic of autochthonous (local) governance should be prioritised to enhance compliance with and participation in regional and international environmental frameworks. However, achieving this requires the decolonisation of the tools and practices through which communities are asked to engage, particularly where obstructive neocolonial logics persist within those frameworks. It also requires addressing the knowledge valve problem; that the regional and international frameworks requiring compliance do not simply extract alternative knowledge where it can be made to fit existing systems, but are instead forced to reckon with these systems by transforming to fit Indigenous local systems.

This is vital for improving regional cooperation on issues such as climate change adaptation. Given the justified reluctance of many nations and communities whose precolonial ontologies remain active to participate in externally imposed systems, decolonial approaches offer a path towards more inclusive and participatory governance: both key tenets of *good governance*. From the standpoint that neocolonial forms of governance obstruct policy development and implementation, and that decolonial governance enhances inclusivity, participation, and thus policy effectiveness, there are compelling reasons to argue that decolonial governance *is* good governance. Moreover, decolonial governance offers a more sustainable form of environmental governance because it is morally accountable to more-than-human custodians of land and sea, an imperative embodied in concepts such as *vanua* in Fiji and *lulik* in Timor-Leste.

In conclusion, it is important to distinguish between *Indigenous governance*, *decolonial governance*, and what might be termed *supracolonial governance*. Indigenous governance is not necessarily decolonial: where it reproduces colonial hierarchies or remains bound to state frameworks, it may reinforce rather than unsettle colonial logics of power. Yet, not all hybrid systems should be dismissed as co-opted. In many cases – such as in the evolving relationship between iTaukei institutions and the Fijian state – hybrid forms endure not as residues of colonialism, but as *supracolonial* configurations: systems that have learned to inhabit, repurpose, and ultimately transcend the colonial state's apparatus. Supracolonial governance thus describes a process through which Indigenous forms thrive *through* and *beyond* colonial infrastructures, using them to generate new modes of authority, legitimacy, and relationality. Rather than a mimicry of colonial forms, supra-colonial governance is an emergent mode that both inhabits and exceeds colonial legacies.

At the same time, neither Indigenous nor supracolonial governance is automatically synonymous with *good governance*. Principles of participation, equity, and inclusivity remain essential tests of governance quality. In Fiji, for example, women's political and customary representation remains limited: only 9.1% of parliamentary seats are held by women (United Nations Women, 2023) and gender-based violence remains among the highest in the Pacific, with roughly 69% of iTaukei women reporting physical or sexual violence in their lifetimes (United Nations Population Fund, 2022). Women's participation on state-owned boards remains around 20%, with few occupying senior roles (Fiji Women's Rights Movement, 2023). For Indigenous or supracolonial governance to be considered *good governance*, the institutions of governance must effectively guarantee human and non-human rights.

They must not only unsettle external colonial hierarchies but also address internal inequities – gendered, generational, or class-based – that limit participation. In this sense, the decolonial promise of governance lies not merely in the assertion of indigeneity, but in the continual transformation of power relations that make governance genuinely inclusive, relational, and accountable to all human and more-than-human custodians.

Notes

1 Anthropologist Sandra Pannell (2011) also writes of the Timorese using forests as refuges to escape colonial violence.
2 The etymology of the prefix 'de' means 'off' or 'from.' For example, decaffeination is a process by which caffeine has been taken *from* the coffee and decapitation is a process by which the head is taken *from* the body.
3 Agenda, 3rd Regional Exchange on Climate Change Adaptation, Dili, Timor-Leste, 29th–31st May, 2013.
4 CTI-CFF "Summary Report: 3rd Regional Exchange on Climate Change Adaptation Dili, Timor-Leste, 29–31 May 2013" p. 8.
5 Country is an English translation of a concept from Aboriginal Australia that refers to the traditional lands and seas over which family groups have custodianship. It is likened to a living being with which aboriginal people have reciprocal custodial relations.
6 For example, *The iTaukei Affairs Board Annual Report 2015–2016* highlights the importance of empowering the *Vanua* through initiatives that promote good governance, leadership, and socio-economic development (p. 4). It also emphasises the need to preserve the *Vanua* by safeguarding traditional knowledge, cultural heritage, and leadership roles, ensuring that they are integrated into modern decision-making processes and development plans (p. 12).

References

Burns, V. 2019, "Traditional Water Management as an Adaptive Subsistence Practice: A Case Study from Coastal Timor-Leste", in *Indigenous peoples and climate* change*: Emerging research on traditional knowledge and livelihoods*, eds. A. Ahearn & R. Kumar Dhir, International Labour Organization, Geneva, pp. 21–33.

Burns, V. 2019, "Oceanic Embodiments: Living ENSO Events in Coastal Timor-Leste", *Political Geography,* vol. 70, pp. 102–116.

Burns, V. 2022, "Analysis of Ocean Ontologies in Three Frameworks: A Study of Law of the Sea Discourse", *Environment and Planning. E, Nature and Space (Print),* vol. 6, no. 2, pp. 1138–1163.

Brennan, A. & Norva, Y.S.L. 2021, "Environmental Ethics: Pathologies of Environmental Crisis – Theories and Empirical Research", in *Stanford encyclopedia of philosophy*, eds. E. N. Zalta & U. Nodelman (eds.). Available: https://plato.stanford.edu/entries/ethics-environmental/theories-research.html

Carter, P. 2018, *Decolonising governance,* 1st edn, Routledge Taylor and Francis Group, London; New York.

Collins, Y.A. 2024, *Forests of refuge,* 1st edn, University of California Press, Berkeley.

Escobar, A. 2011, "Sustainability: Design for the Pluriverse", *Development (Society for International Development),* vol. 54, no. 2, pp. 137–140.

Escobar, A. 1995. *Encountering development: The making and unmaking of the third world*, Princeton University Press, Princeton, NJ.
Escobar, A. 2018, *Designs for the pluriverse,* 1st edn, Duke University Press, Durham; London.
Fanon, F. 1967, *The wretched of the earth,* Penguin, Harmondsworth.
Fiji Women's Rights Movement, *Unmet Promises: Women on Boards in Fiji Still Well Below 30%*. Available: https://www.fwrm.org.fj/news/media-releases/unmet-promises-women-on-boards-in-fiji-still-well-below-30-8-06-2023 [2025, 9th October].
Garba, T. & Sorentino, S. 2020, "Slavery is a Metaphor: A Critical Commentary on Eve Tuck and K. Wayne Yang's "Decolonization is Not a Metaphor"", *Antipode,* vol. 52, no. 3, pp. 764–782.
Hage, G. 2017, *Is racism an environmental threat?* 1st edn, Polity, Cambridge.
Hage, G. 2008, *Remember, the roots will be with you always,* Nationwide News Pty Ltd, Canberra, A.C.T.
Jolly, M. 1992, "Specters of Inauthenticity", *The Contemporary Pacific,* vol. 4, no. 1, pp. 49–72.
Lewis, D. & Mosse, D. 2006, *Development brokers and translators: The ethnography of aid agencies,* Kumarian Press, West Hartford.
Merry, S.E. 1988, "Legal Pluralism", *Law & Society Review,* vol. 22, no. 5, pp. 869–896.
Mignolo, W.D. 2009, "Epistemic Disobedience, Independent Thought and Decolonial Freedom", *Theory, Culture & Society*, vol. 26, no. 7, pp. 159–181.
Mignolo, W.D. 2011, *The Darker side of western modernity,* 1st edn, Duke University Press, Durham; London.
Mignolo, W.D. & Walsh, C.E. 2018, *On decoloniality,* 1st edn, Duke University Press, Durham; London.
Muecke, S. & Roe, P. 2021, *The children's country: Creation of a goolarabooloo future in North-West Australia*, Rowman & Littlefield, London.
Murrey, A.a .D., P. 2023, *Learning disobedience,* Pluto Press, London.
Nayacakalou, R.R. 1975, *Leadership in Fiji,* Oxford University Press, Melbourne; London.
Ndlovu-Gatsheni, S.J. 2015, "Decoloniality as the Future of Africa", *History Compass,* vol. 13, no. 10, pp. 485–496.
Ndlovu-Gatsheni, S.J. 2016, *Theœ decolonial Mandela,* 1st edn, Berghahn Books, New York; Oxford.
Ndlovu-Gatsheni, S.J. 2021, "The Cognitive Empire, Politics of Knowledge and African Intellectual Productions: Reflections on Struggles for Epistemic Freedom and Resurgence of Decolonisation in the Twenty-First Century", *Third World Quarterly,* vol. 42, no. 5, pp. 882–901.
OED. 2025, *supra-, prefix,* Oxford University Press, Oxford.
Pahuja, S. 2011, *Decolonising international law: Development, economic growth and the politics of universality,* Cambridge University Press, Cambridge [England].
Pannell, S. 2011, "Struggling Geographies: Rethinking Livelihood and Locality in Timor-Leste", in *Land and life in Timor Leste: Ethnographic essays,* eds. A. McWilliam & E. Traube, ANU Press, Canberra, pp. 217–240.
Peters, K. 2020, "The Territories of Governance: Unpacking the Ontologies and Geophilosophies of Fixed to Flexible Ocean Management, and beyond", *Philosophical Transactions Royal Society B.,* vol. 375, pp. 1–10.
Pietri, D.M., Stevenson, T.C. & Christie, P. 2015, "The Coral Triangle Initiative and Regional Exchanges: Strengthening Capacity through a Regional Learning Network", *Global Environmental Change,* vol. 33, pp. 165–176.

Polanyi, K. 2024, *The great transformation: The political and economic origins of our time,* New edn, Penguin Classics/Penguin, London.
Quijano, A. 2000, "Coloniality of Power and Eurocentrism in Latin America", *International Sociology,* vol. 15, no. 2, pp. 215–232.
Quijano, A., Mignolo, W.D., Segato, R.L. & Walsh, C.E. 2024, *Aníbal Quijano,* 1st edn, Duke University Press, Durham.
Smith, L.T. 2012, *Decolonizing methodologies,* 2nd edn, Zed Books, London; New York.
Smith, L.T., Tuck, E. & Yang, K.W. 2018, *Indigenous and decolonizing studies in education,* 1st edn, Routledge, London.
Spivak, G.C. 1988, "Can the subaltern speak?" in *Marxism and the interpretation of culture,* ed C.a. G. Nelson Lawrence, Macmillan, London, pp. 271–316.
Sultana, F. 2025, *Confronting climate coloniality: Decolonizing pathways for climate justice: pbk,* Routledge, London.
Sultana, F. 2022, "The Unbearable Heaviness of Climate Coloniality", *Political Geography,* vol. 99, p. 102638.
Thaman, K.H. 2003, "Decolonizing Pacific Studies: Indigenous Perspectives, Knowledge, and Wisdom in Higher Education", *The Contemporary Pacific,* vol. 15, no. 1, pp. 1–17.
Thaman, K.H. 2019, "Sustainable Development for Whom? A View from Oceania", *International Education Journal,* vol. 18, no. 3, p. 1.
Tuck, E. & Wayne Yang, K. 2012, "Decolonisation is Not a Metaphor", Decolonization*: Indigeneity, Education & Society,* vol. 1, no. 1, pp. 1–40.
UN Women 2023, *Country Data Fiji*. Available: https://data.unwomen.org/country/fiji [2025, 9th October].
United Nations Development Program, Asia and the Pacific 2020, Papua New Guinea*: Ensuring Women Have a Say - Putting Gender at the Heart of Free, Prior and Informed Consent*. Available: https://Papua New Guinea: Ensuring Women Have a Say – Putting Gender at the Heart of Free, Prior and Informed Consent [2025, 9th October].
United Nations Economic and Social Commission for Asia and the Pacific , *What is Good Governance*. Available: https://www.unescap.org/sites/default/files/good-governance.pdf [2025, 17th September].
Ward, R.G. 1995, *Land, custom and practice in the South Pacific,* 1. publ. edn, Cambridge University Press, Cambridge [u.a.].

PART II

Challenges to good governance

6

AFTER THREE DECADES OF SUSTAINABLE DEVELOPMENT, HOW DO PUBLIC AND PRIVATE ORGANISATIONS IN THAILAND GOVERN SUSTAINABILITY WITHIN THEIR ORGANISATION?

Molly Wimonmat Srichamroen

Background and introduction

'Good governance' has been a key topic of discussion in both the public and private sectors in Thailand. For the public sector, as per Addink's (2019) definition, good governance reflects how well a government functions to meet the needs of its people. It includes the systems, procedures, and actions to address public concerns, manage resources responsibly, and legitimately exercise authority within a community. In the Modern State, good governance is more than just strengthening the rule of law and democracy; it also crucially involves government accountability and responsiveness (Addink, 2019). The government uses public money for societal development, so it is natural for the goodness of governance of public organisations to be judged in connection with development outcomes. However, the processes used to obtain those outcomes also matter for good governance, given that the existence and activities of the public sector are fundamentally *for* the public (Osborne, Radnor, and Nasi, 2013). Given that the source of a public organisation's budget comes from the public, ethical practice, transparency, accountability, and stakeholder engagement are all expected.

As outlined by The United Nations Economic and Social Commission for Asia and the Pacific (UNESCAP), the principles of good governance are: participatory, consensus-oriented, accountable, transparent, responsive, effective and efficient, equitable, inclusive, and follow the rule of law (UNESCAP, 2009). UNESCAP explicitly states that these principles should be employed in both the public and private sectors. As opposed to the public sector, whose monetary goal is to ensure public money is spent efficiently (Mandl, Dierx,

DOI: 10.4324/9781003467694-8

and Ilzkovitz, 2008; UNESCAP, 2009), in the case of private organisations, profitability is essential for survival and growth. Nevertheless, good governance in the private sector also includes ethical practices, employee well-being, accountability, customer focus, and societal contributions. Stakeholder expectations mandate responsibility for internal management, the operating environment, and all affected parties.

In its sustainable development efforts, Thailand has been significantly influenced by the United Nations' Sustainable Development Goals (SDGs). The country is adapting these goals to empower local governments and communities and work towards reducing inequality. In parallel, national long-term and short-term strategies, development roadmaps, and ministry-level strategies explicitly state that the country's objectives are linked to all SDGs. The SDGs are also increasingly being incorporated into the strategies of private sector organisations, which play a vital role in the country's economic development through their productivity. Given their macro-level scope, the SDGs can be challenging for public and private organisations to directly apply. Consequently, at the organisational level, the focus shifts to daily management and processes that contribute to sustainable development, collectively referred to as sustainability management.

Sustainability management means a process to ensure that things are maintained and not getting worse for future generations, while also improving people's lives by alleviating detrimental conditions. It combines the notion of sustainability, which is about keeping things going at a constant level, with development, which is about improving people's lives by reducing inequality, poverty, unemployment, and environmental damage. It recognises the social, environmental, and economic aspects of sustainability while emphasising also the importance of governance processes for achieving progress in all of these areas.

Therefore, despite their different missions, core good governance and sustainability principles must guide all organisations' internal management in both the public and the private sector. This qualitative study explores whether organisations across the public and private sectors in Thailand share a common understanding of sustainability management and good governance practices. The argument is structured as follows. The next section will outline Thailand's current sustainability landscape at both the macro level and organisational level across both the private and public spheres. This chapter's methodology will then be summarised, with a focus on the framework used for analysing self-governance for sustainability in the studied organisations. The findings from interviews with public and private sector managers will then be presented and discussed, in combination with secondary sources (organisation websites, reports, and documents). This chapter's conclusion will be a reflection on the differences identified and lessons to improve sustainability governance at the organisational level in Thailand.

Sustainable development in Thailand

At the national level

Thailand first addressed sustainability at the macro level in its 7th National Economic and Social Development Plan (1992–1996). Current efforts are guided by the 13th Plan (2023–2027) and the 20-Year National Strategy (2018–2037), which serve as key frameworks for SDG action (Government Public Relations Department, 2016). Chaired by the Prime Minister, the National Committee for Sustainable Development (CSD) leads policy and fosters public, private, and civil partnerships.

These national strategies are rooted in the Sufficiency Economy Philosophy (SEP), which emphasises moderation, reasonableness, integrity, and prudence, as envisioned by the late King Bhumibol Adulyadej (Ministry of Foreign Affairs of Thailand, 2017a). SEP promotes holistic, qualitative development, emphasising social safety nets and social capital (Mongsawad, 2010). This paradigm addresses challenges in institutional development, human capital, the natural environment, and the role of the government (Mongsawad, 2010; Mongsawad and Thongpakde, 2016). Together with the notion of sustainable development, SEP is the foundation for Thailand's constitution (Royal Thai Embassy, 2024), and its influence also extends to the private sector, guiding ethical operations, stakeholder consideration, good governance, social responsibility, transparency, and risk management (Mongsawad, 2010; Suttipun, 2018). There is a clear overlap between good governance and SEP principles in Thailand, in both its public and private sectors.

While the country annually reports on SDGs implementation as part of Voluntary National Reviews (Ministry of Foreign Affairs of Thailand, 2017b), actual impact assessment is challenging. These reports, published by the Ministry of Foreign Affairs of Thailand detail actions and outputs, not outcomes; for instance, they show low-income benefit registrations and high Universal Health Coverage, but not poverty reduction or actual healthcare accessibility (Ministry of Foreign Affairs of Thailand, 2017b). Another example is related to female empowerment. The 2022–2023 report detailed the Thai Women Empowerment Fund's establishment, member organisations, and government funding requests, but lacked information on money utilisation or project impact (Ministry of Foreign Affairs of Thailand, 2024). While national statistics like GDP show development improvements, others present a stark contrast: the Bank of Thailand (2023) reported rapidly increasing household debt, posing a potential threat to the country's financial stability. Furthermore, the number of homeless individuals has been increasing in Bangkok and other provinces as a result of higher unemployment (Pattaya Mail, 2022). This raises the question of the role of public and private institutions in driving national development.

Despite Thailand's development policy explicitly incorporating good governance (e.g. effectiveness and efficiency) and SEP principles (e.g. ethics and integrity) at all levels, it remains unclear to what extent these are genuinely understood and implemented in the daily operations of organisations within the country. The extent to which the national goal of sustainable development is adopted and operationalised below state level is a complex challenge, which we turn to next.

In public and private organisations

From an organisational perspective, sustainable development means long-term viability of the organisation through positive contributions to society and the environment. In achieving this goal, the organisation needs enablers of good governance, such as stakeholder engagement or long-term value creation, which must be fostered and continuously enhanced. The organisation-level management effort for sustainable development thus goes beyond environmental and financial compliance, or philanthropic gestures (United Nations, 2015; Oertwig et al., 2017; Busco, Riccaboni, and Nasi, 2017).

Ethical codes are vital for sustainable development in both the public and private sectors. However, these sectors operate differently, and sustainability management in the public sector is notably challenging and has progressed slowly, despite growing awareness of its importance (Figueira et al., 2018; Domingues et al., 2017). Public organisations bear greater societal responsibilities due to their public mission and reliance on public funds to provide welfare and improve people's quality of life. Consequently, they are financially and ethically obligated to serve society. Public sector entities face limited competition, often being the sole provider of essential goods and services (Kraft and Furlong, 2019). In Thailand, some services, like electricity, intercity rail transport, and water, are exclusively provided by state enterprises. The advantages of this monopoly are the affordability and accessibility of these services at a low cost for the public. The disadvantage in such a system is that the quality of service is up to the provider, making it provider-dominant (Heimler, 2000). While not profit-driven like the private sector, the public sector's institutional robustness is crucial; its stability is vital for national safety against uncertainties.

Public organisations, like ministries, base their sustainability strategies on national plans such as Thailand's 12th Economic and Social Development Plan and the 20-Year National Strategy (2018–2037), which aim for 'stability, prosperity, and sustainability.' These plans are guided by the SEP, emphasising a balanced economic system, public health, economic well-being, and reduced inequalities (NESDC, 2019). However, aligning strategies across different levels can be complex. Effective implementation and achieving national sustainability goals require public leaders to deeply understand sustainability in practice.

Private sector organisations are driven by unique stakeholder expectations and growth ambitions, shaping their management approaches. As Open System Theory suggests, the external environment dictates strategies and mindsets (Scott and Davis, 2007). Companies constantly compete for revenue, requiring continuous improvement to gain or maintain market share. A strive to exist and thrive in the market is a basis for private organisations' survival. Beyond employees and customers, crucial stakeholders include investors and shareholders, whose returns depend on company performance. From a sustainability perspective, private organisations' stakeholders encompass immediate communities and the broader society impacted by their operations, pressuring them towards social responsibility beyond competitiveness.

Methodology

Aim of this study

The main aim of this study is to explore how, public and private organisations in Thailand incorporate sustainability within their day-to-day operations, in order to assess whether their governance could be classed as 'good.' To help us understand to what extent organisations foster good governance through their sustainability practices, Galbraith's (2002) Star Model framework is employed here. The model includes five elements which contribute to the internal sustainability of an organisation: strategy, structure, people, processes, and rewards, all of which will now be explained in detail.

Strategy dictates an organisation's overall direction, products, presence, and activities (Galbraith, 2020). Bryson and George (2020) define it as a concrete plan for achieving goals and creating public value by aligning aspirations and capabilities. Strategy is reflected in an organisation's own mission, values, and vision (Bryson and George, 2020; Bryson, 2018). For private companies, strategy also includes market engagement and profit generation. For the purposes of this study, it is important to assess whether sustainability is explicitly part of Thai organisations' strategy, both on a general level and also in terms of explicit references to the environmental, social, and governance pillars of sustainability.

According to Kates and Galbraith (2010), structure determines where authority resides and how decisions are made. An organisation's structure can be presented in an organisational chart, which shows the organisation's hierarchy and reporting relations, and highlights key functional areas or what the organisation sees as a priority. In organisational management literature, structure also refers to the organisational framework, including the hierarchical relationships and lines of authority within a company (Kaplan, 2005). This study explores how sustainability management is incorporated in the organisational structure of each organisation, assessing whether there are designated units or departments within the organisation responsible for sustainable development, and an explicit reporting line of command.

'People' refers to the human resources aspect of an organisation, including talent management and key functions such as training and recruitment. They are an organisation's workforce, composed of individuals with diverse backgrounds, including varying levels of knowledge, experience, abilities, and training. The organisation's strategy and structure play a role in determining what specific skills, mindsets, or values people need in order to achieve the organisation's goals. For the purposes of this study, we are particularly interested in whether organisations explicitly include training for their employees to develop their sustainability skills.

Processes are the formal and informal management procedures within an organisation, encompassing areas such as information transmission, performance evaluation, planning, budgeting, and resource allocation (Kaplan, 2005; Kates and Galbraith, 2010; Galbraith, 2020). This study assesses whether sustainability-supporting processes are in place in Thai organisations.

Rewards focus on the employees. They are designed to align employees' personal goals with the organisation's objectives. The alignment plays a motivating role for the employees to perform at their highest level (Galbraith, 2020). This study explores whether rewards in Thai organisations are designed to incentivise employees to be more sustainable.

Research design: case study analysis

To gain an in-depth and detailed understanding of sustainability in private and public organisations in Thailand and its interplay with good governance, this study employs a qualitative research design, consisting of a comparative case study of three Thai organisations: a large private organisation (Kasikorn Bank [KBank]), a small private organisation (Digio), and a public organisation, The Office of National Economic and Social Development Council (NESDC). These were sampled purposively, for the reasons detailed below.

Kasikorn Bank

KBank (31,700 employees) is the first bank in Thailand to have been listed in the S&P Dow Jones Sustainability Indices and has remained on the list since 2016, which reflects its continuous sustainability management efforts. KBank is subject to stringent sustainability requirements, such as the Thai Securities and Exchange Commission's rules on good corporate governance and developing sustainable business practices and the Stock Exchange of Thailand's (SET) environmental, social, and governance sustainability rating assessments. Accordingly, the bank needs to develop and disclose their performance based on sustainability metrics. Furthermore, it must meet comprehensive reporting requirements, following the SET's guidelines for listed companies, the globally recognised Global Reporting Initiative (GRI) standard, and align its performance with the SDGs (KBank, 2025).

While many Thai companies listed on the stock exchange likely face similar sustainability obligations, KBank was chosen for this study due to its exceptional global sustainability performance. KBank stands out as the sole Thai bank that has consistently ranked within the top 1–5% of the Corporate Sustainability Assessment (CSA), a global assessment by S&P Global, for three consecutive years in 2022–2024.

Digio

A private fintech company (156 employees), Digio offers a range of cashless financial solutions to businesses, from small enterprises to large corporations. Digio's sustainability practices are not subject to evaluation by any ratings or indices. It was chosen for the study as a case of a small business obligated to follow the sustainability practices of its outsourcing partner, KBank. Since Digio's core business is also in providing services in finance, it offers insight into how a smaller organisation with few resources and no formal institutional sustainability reporting requirements approaches sustainability management and development at organisational level.

By analysing sustainability management at KBank and Digio, we can gain valuable knowledge about sustainability management practices in private companies with varying levels of investment, business scale, and organisational size in comparison to government organisations.

The office of National Economic and Social Development Council

NESDC is a government organisation (600 government officers employed) whose core mission is developing the country's national strategies and plans, including the National Economic and Social Development Plans where the concept of sustainability was first introduced in Thailand. The organisation conducts studies to recommend national policies and strategies in the sustainability arena. It also facilitates coordination between the government, relevant agencies, and state enterprises in the planning and implementation of development programmes and projects.

Data collection methods

In-depth semi-structured interviews and documentary analysis were employed to gather data and information related to sustainability within the studied organisations. For the interviews, ethical approval was received from the Institutional Review Board of Chulalongkorn University. The people interviewed were individuals whose responsibilities include sustainability management within the selected organisations, and were initially approached via email. They were given the opportunity to ask questions about the study, and signed a consent form prior to being interviewed. Participants were also informed that they had the right to withdraw from the study within two weeks of the interview if they changed their mind. Interviews took place in person during September–December 2024, lasted 60–90 minutes each, were recorded, and were conducted in Thai and translated into English subsequently. To protect participants' anonymity, only the organisation will be mentioned in the analysis that follows, without any references to personally identifiable data. Interview questions were closely based on the Star model framework, covering all five dimensions (strategy, structure, people, processes, and rewards). These were explored both in general within each organisation and also specifically in relation to whether they contribute to sustainability.

Data collection also involved secondary data: organisational documents and publications (annual reports, sustainability reports, financial reports), as well as the organisations' websites, which include information on the organisations' vision, mission, and strategy, employee demographic data, management structures and hierarchy, policies, and news about the organisation.

Data analysis

The qualitative data from interviews, documents, and websites were analysed thematically, with the five dimensions of the Star Model as themes, in connection with the environmental, social, and governance pillars of sustainability. NVivo was used to organise the data and facilitate thematic analysis.

Findings

A significant contrast emerged in the analysis of sustainability management practices especially between the government organisation and the bank examined in this study, with the small business sharing similarities with one or the other across the different aspects assessed. Table 6.1 illustrates the main findings, which are then explained in detail in the following subsections. It shows whether each of the three organisations studied incorporates sustainability considerations across the five dimensions of the Star Model, both in general terms and separately for the environmental, social, and governance pillars of sustainable development. The columns are sustainability aspects and the rows are the Star Model dimensions. The symbols in the table mean the following: / The organisation has a formal requirement set up for sustainability management activities and/or sustainability is internally and systematically monitored and evaluated; O The organisation implements the sustainability

TABLE 6.1 Assessment of sustainability practices (based on the Star model framework) in the studied organisations

Star Model organisation elements and related sustainability management activities	*Overall*			*Environment*			*Social*			*Governance*		
	KBank	Digio	NESDC	KBank	Digio	NESDC	KBank	Digio	NESDC	KBank	Digio	NESDC
Strategy												
Sustainable development is part of the organisation's vision and strategy	/	x	o	/	x	x	/	x	x	/	o	o
Sustainability strategic goals are set	/	o	o	/	x	x	/	x	/	/	o	x
Sustainability strategic performance is audited by third-party	/	o	x	/	x	x	/	o	x	/	o	o
Sustainability strategic performance is disclosed	/	x	x	/	x	x	/	x	x	/	x	x
Structure												
Designated organisational-level department/unit/ staff responsible for sustainability management	/	o	x	/	o	o	/	o	o	/	/	o

(Continued)

TABLE 6.1 (Continued)

Star Model organisation elements and related sustainability management activities	*Overall*			*Environment*			*Social*			*Governance*		
Disclosure of organisational sustainability management structure	/	x	x	/	x	x	/	x	x	/	x	x
People												
Mandatory training/ programme for employees to develop sustainability skills	/	o	x	/	x	x	/	x	o	/	x	x
Employee performance evaluation includes their sustainability skills development	/	o	o	/	x	x	/	x	x	/	x	x
Management create a sustainability culture	/	o	x	/	o	x	/	x	x	/	o	x
Processes												
Sustainability-supporting systems are in place (data collection, training, evaluation)	/	o	o	/	o	x	/	o	o	/	x	o
Sustainability-supporting systems performance is evaluated	/	x	x	/	x	x	/	x	x	/	x	x

TABLE 6.1 (Continued)

Star Model organisation elements and related sustainability management activities	*Overall*			*Environment*			*Social*			*Governance*		
HR management includes ESG aspects	/	o	o	/	o	o	/	o	o	/	o	o
Rewards												
The organisation offers rewards for sustainability-related campaigns or behaviours	/	x	x	/	x	x	/	/	/	/	x	x
The organisation offers employment benefits or rewards relating to sustainability development (well-being, environment, or good governance)	/	/	o	/	/	x	/	/	/	/	/	/

activity but without internal systematic monitoring and/or evaluation, or only in some parts of the organisation; X The organisation does not implement the activity.

Strategy

Strategy planning reflects both similarities and differences across the public and private organisations studied. Table 6.2 shows the studied organisations' vision, mission, and core values (key aspects of strategy), drawn from each organisation's website. Where the website was not available in English, this has been translated into English by the author.

In terms of similarities, all three organisations consider and analyse their internal and external environment, including potential risks and opportunities (SWOT analysis), a well-known tool in strategy development. Executives engage in developing strategy and performance targets. In private organisations, this also comes with transparency in third-party auditing and public disclosure. For the smaller private company and for the government organisation, a sustainability strategy for the organisation is not specifically developed. Nonetheless, it is important to note that the public organisation's sustainability strategy is interpreted and primarily oriented towards public benefit at the national level, rather than internal sustainability. This leads to less emphasis on internal processes as elements of good governance. By contrast, the large private organisation views sustainability strategy and performance as a competitive advantage, benefiting both their business growth and fulfilling their social responsibilities.

With embedded commitment to sustainable development in the organisation's vision and mission, KBank's sustainability strategy, targets, and performance evaluations are included in every business unit's operation. The bank's Executives and Board Members take part in planning the organisation's sustainability strategy and their individual performances are based on the outcomes of the strategy. All their sustainability strategy and performance evaluations are third-party verified, reported and disclosed publicly. Digio, on the other hand, does not have an explicit organisational level sustainability strategy, although implicitly it does prioritise social sustainability in the form of building trust and taking responsibility for clients and employees. Business risks are assessed, emphasising Digio's first core value, trust. This is especially true for data protection and security. A zero-breach target is set to ensure clients' and employees' data security and protection. Digio's performance is audited and verified by a third-party organisation. The owner of the company, together with Department Managers, reviews the previous year's performance together and plans the next year's business strategy (including sustainable development considerations), but sustainability performance as such is not publicly disclosed.

TABLE 6.2 Vision, mission, and core values in the studied organisations

Strategy	*Kasikorn Bank*	*Digio*	*NESDC*
Vision	'KASIKORN BANK aims to be the most innovative, proactive, and customer centric financial institution, delivering world class financial services and sustainable value for all stakeholders by harmoniously combining technology and talent.' (KBank, n.d.)	'Payment Simplified: We create innovative payment solutions that revolutionise the way you do business.' (Digio, 2022)	'Central planning agency responsible for strategy formulation towards balanced and sustainable development while upholding national interests and keeping up with unexpected changes with high efficiency.' (NESDC, n.d.)
Mission	'Delivering Sustainable Value to Stakeholders: Customers: Make our customers successful in their life and business Shareholders: Generate sustainable total returns Employees: Provide learning and leadership opportunities Regulators: Co-Create the future of finance Society: Thrive for better quality of living.' (KBank n.d.)	'We work each and every day to help people grow their business.' (Digio, 2021)	'Formulate and integrate national development plans; Coordinate and drive national development Evaluate and assess the progress of national development Be a knowledge and information centre for national development and Be an efficient internal management department.' (NESDC, n.d.)
Core values	'Having the customer at heart, collaboration, agility, innovation, and integrity.' (KBank, n.d.)	'Integrity: Building trust from integrity. Value Added: Creating value for ourselves, the company, and customers. Efficiency: Make it work, make it right, make it fast. Responsibility: Taking ownership and responsibility. Respect: Be respectful, treat others the way you want to be treated.' (Digio, 2021)	'Dedicated to excellence; social and public responsibility; good governance; human resources as the most valuable asset.' (NESDC, n.d.)

Despite having developed multiple national strategies with a focus on sustainable development, at organisation level, NESDC lacks an explicit organisational sustainability management strategy. Its focus remains on co-ordinating national SDG implementation. While internal and external risks are assessed for national development, no overall organisational sustainability strategy or operational report is publicly disclosed. Interestingly, while NESDC is the only organisation which includes good governance explicitly amongst its core values, it does not clearly define what is meant by this. However, a 'message from the Secretary-General' article clarifies that for the NESDC's own conduct, good governance encompasses 'responsiveness, reliability, efficiency, engagement, agility, accountability, lean operations, and learning as a team' (NESDC, n.d.). It is worth noting that while this document is available online, it is not discoverable through a simple manual search on the NESDC's website; this suggests that good governance is assumed to be a generally agreed-upon term without the need for a definition.

Structure

The large private organisation, KBank, integrates sustainability through a Board-supervised Corporate Sustainability Integration Department, which consolidates sustainability data reporting across all units. A President-chaired Climate Pillar Working Group manages GHG emissions from both operations and financed portfolios, developing action plans aligned with their Net Zero Commitment (Kasikorn Bank, 2024). KBank embeds sustainability management in all business departments as the following interview excerpt suggests.

> We do not only have one single party committing to sustainability practices and assessing sustainability risks [in social, environment, and governance aspects], but we have multiple parties dispersed all over the organisation with these aspects included in their operations, from Corporate to Innovations. We also have a business unit overseeing GHG emission reduction for our clients and ESG risk assessment units overseeing the bank's own sustainability risks and our clients' ESG portfolios. There is a unit overseeing the integration of the overarching sustainability practices. We also have K-Climate specifically overseeing just the environmental aspect.
>
> *(Interview, author's own translation, Kasikorn Bank Interview Participant, Sept 2024)*

Being a smaller business, Digio designates individual staff, rather than a department, to be responsible for the organisation's social, environmental, and governance pillars of sustainability. As the company works with big corporate clients, they are obligated to collect some carbon data such as emissions (Scopes 1–2) and report them to their clients. Designated staff is responsible for data collection. A clear responsibility regarding sustainability management is assigned and communicated. The Head of the Organisation who is the owner of the company (owning 70% of the shares) is unofficially responsible for sustainability management. The Head of Organisation directly engages with clients with strong sustainability awareness and standards, including vendor assessments. This direct interaction allows management to gain valuable insights and subsequently drive organisational awareness and set higher sustainability performance expectations for all employees.

In the public sector, the NESDC does not have a dedicated sustainability department. Responsibility for internal sustainability is fragmented across existing functions and largely unrecognised as part of broader sustainability initiatives. While implicit practices include employee well-being, data security, and anti-corruption, only environmental sustainability is officially recognised, overseen by the Secretary-General's Office, which unofficially leads all sustainability management. Furthermore, strictly following bureaucratic regulations causes risks in multiple aspects. The interviewee mentioned blind procurement, suggesting that it may allow corruption to take place inadvertently.

> Our E-bidding system is supposed to enhance the transparency of the procurement process. We don't know who the bidders are. By the time the bidding is finalised, the winning vendor signs the contract. If we happen to find out later that they have a relationship or connection with someone within our office [who influences the decision in the work under contract], we are not able to do anything, because the contract will have already been finalised.
>
> *(Interview, author's own translation, NESDC Interview Participant, Dec 2024)*

Furthermore, as a consequence of not having a specific unit responsible for internal sustainability management, limited relevant sustainability data such as employee health and safety, GHG emissions, data on the reasons for complaints and grievances, and recycling data is collected as part of the NESDC's operations.

People

At KBank, sustainability expertise extends beyond the designated sustainability departments mentioned above. Employees in areas such as Human Resources and the Corporate Governance Committee also possess crucial competencies, especially in governance and social sustainability, enabling them to share technical knowledge bank-wide. Furthermore, KBank embeds sustainability throughout its operations by fostering a 'Green DNA' culture. Through evaluating and publicly disclosing social, environmental, and governance performance, the bank encourages employees to prioritise both business growth and sustainable personal practices such as bringing their own reusable coffee mug to a café, carrying a tote bag, and taking a SkyTrain to work instead of driving.

Digio recognises that both the company and its employees are still in a learning phase when it comes to developing and sustainability skills. However, given that some data, particularly in the social and environmental areas, is requested by major clients, this presents an opportunity for the company to learn from client expectations and improve its own and its employees' sustainability performance. Additionally, with an organisational culture based on trust, Digio adheres to its mission of helping its stakeholders, as demonstrated by the following interview excerpt.

> If we detect an emerging business risk, we don't care about our profit, we care more about clients. There was a case where our system malfunctioned and our client lost money; in that situation, we repaid them. We take responsibility and it gains us their trust in return.
>
> *(Interview, author's own translation, Digio Interview Participant, Oct 2024)*

Whereas trust and integrity are already a part of the organisation's sustainability strategy, the interview participant mentioned that developing a reliable secured payment system has only become a key principle more recently. There are no shared sustainability skills in other areas.

The NESDC's sustainability knowledge, skills, and training landscape is complex and fragmented. Many functions implicitly contribute to sustainability, yet this is not explicitly acknowledged. Management performance in terms of transparency and accountability is not disclosed. Human resources evaluations focus solely on activity completion. Consequently, current sustainability awareness and skill development make high-level organisational sustainability expertise unlikely among all officers. It is difficult to identify a shared value for sustainability within NESDC as no specific sustainability practice stands out as receiving particular attention or emphasis. However, the NESDC vision, mission, and core values reflect its broad contributions to Thai society, with employees being generally public-minded rather than focusing on smaller sustainability initiatives at the organisational level.

Processes

KBank integrates its sustainability strategy across departments and units, led by the Corporate Sustainability Integration Department, and supported by central functions like IT and HR. Environmental policies, including conservation goals and resource use reduction guidelines, contribute to sustainability governance. Transparent governance processes, such as the Code of Conduct communication, director training, whistleblowing procedures, and governance risk management, are publicly disclosed, along with annual performance evaluations.

Likewise, Digio systematically tracks environmental and social sustainability data, including emissions from operations, transport, waste, and electrical consumption. Data security is continuously monitored and independently verified. Employee diversity data is also collected. The company plans to enhance digital data collection to meet stakeholder needs, as mentioned by the interviewee:

> We plan to develop a system to collect more data as our clients need [environmental, social, and governance] performance data from vendors for their own sustainability management.
>
> *(Interview, author's own translation, Digio Interview Participant, Oct 2024)*

However, it must be noted that Digio does not have digital or training systems in governance specifically in place. Related data and information are not publicly disclosed.

NESDC does not explicitly collect and report data on internal environmental or social sustainability or governance processes. Data security and privacy are tested periodically and frequently by the organisation's internal IT department and by external contractors, and the organisation employs an executive responsible for digital matters, who periodically attends training programmes. However, NESDC's IT system is at high risk, threatening the organisation's sustainability, as mentioned by the following interviewee:

> Being hacked is quite an issue. Hackers attempt to hack us all the time. In fact, government systems are hacked everywhere. I am not sure if we have a target regarding the number of breaches, like no more than a certain number of times a year, but there should not be any breaches. From our monitoring, we see that hackers try to hack our systems all the time, some days more than a thousand times…
>
> *(Interview, author's own translation, NESDC Interview Participant, Dec 2024)*

Another sustainability issue appears to be workload, which affects employees' ability and willingness to attend the various types of training offered by the organisation (first aid, fire safety, CPR):

> We have first aid training, but it is not taken seriously. Most of our academic officers ignore the fire alarms during the drill because they are worried about their unfinished work.
>
> *(Interview, author's own translation, NESDC Interview Participant, Dec 2024)*

There is also a lack of a formal vendor vetting process within the NESDC. Vendor performance is currently self-assessed. Notably, the Public Procurement and Supplies Administration Act of 2017 prohibits officers involved in procurement from having any conflicts of interest with tenderers, but also states that:

> In the case where it appears thereafter that the person having the duty to conduct the procurement under paragraph one has any interest with the tenderer of a proposal or the contractual party in any process of the procurement or is a member of a procurement committee without material prejudice to the procurement, such procurement shall not be impaired.
>
> *(Public Procurement and Supplies Administration Act B.E.2560 (2017), Section 13)*

There is no mention of a systematic vetting process allowing a certain level of risk to be accepted. Given the level of risk involved across the different processes within NESDC, both sustainability and good governance are called into question.

Rewards

KBank's Board of Directors drives a top-down sustainability culture, including 'Green DNA.' The bank's sustainability initiatives include allowing employees to take time off for volunteering, as well as 'financial health' retirement planning sessions. Given that sustainability performance is also integrated into employee evaluations, all levels of staff at KBank are incentivised to develop sustainability skills and demonstrate a commitment to sustainability. In addition, KBank executives (Vice-President level and above) can choose to be provided with a company electric car or to receive a subsidy for a personal electric car.

For Digio and NESDC, sustainability skills are not as much recognised specifically, because these skills may be fractured across different activities and units. Without an explicit focus of the organisation's strategy directly connected to internal sustainability, there is no incentive to seek or develop sustainability skills in particular, any more than other skills. However, all organisations provide sustainability benefits to their employees, such as well-being benefits in the form of free health-check-ups and private health insurance.

Discussion and conclusions

The analysis above revealed important differences between public and private organisations in Thailand when it comes to self-governance for sustainability.

Both KBank and Digio are externally pressured and motivated to develop and improve their sustainability performance by market competition and conditional business transactions. KBank in particular has integrated sustainability management practices throughout the organisation and extends these to its vendors and subsidiaries. NESDC on the other hand does not have this external motivation and pressure. However, NESDC itself is an organisation that creates national plans and strategies that focus on the country's sustainable development; as such, sustainability strategy and practices within NESDC are then more externally oriented, towards public benefit more than the organisation's own performance. This leaves us with an important question for further research: How do public organisations, whose mission directly aligns with national sustainable development aims, recognise and assess their own organisational sustainability performance?

This is crucial because strong sustainability internally is reflective of effective communication, strong teamwork, or impactful stakeholder engagement (Da Silva and Romaro, 2024; Brundiers and Wiek, 2017), all of which then have an impact on external outcomes. In other words, good governance in terms of processes has a direct impact on good governance in terms of outcomes. The unavailability of systems, data, and ways to assess organisational sustainability within NESDC reflects traditional bureaucratic management, which is challenged by the fast-paced national environment which necessitates the development of up-to-date practical national plans. Consequently, it is challenging for the public to assess whether the organisation effectively manages its missions in a sustainable manner and utilises public funds efficiently. The performance and impact of the government's effort in Thailand's sustainable development progress cannot be directly determined, in turn making it difficult to assess the national strategy's adherence to good governance principles like ethics and integrity, as demonstrated also by recent events.

The collapse of the State Audit Office construction site in Thailand during the March 2025 earthquake was followed by allegations of corruption and a lack of accountability on the part of both the public and private organisations involved. Similarly, the RAMA II Road and overpass construction, a project that has dragged on for decades, also suggests a failure to uphold good governance principles such as accountability, reliability, efficiency, responsiveness, and a negligence of sustainability management practices in Thailand. The absence of adequate health and safety measures on the management of this project has led to a toll of over 2,500 construction-related accidents and more than a hundred deaths among both workers and road users (Nation Thailand, 2025).

Decades of sustainable development efforts in Thailand have yielded lessons that can be applied to better align national SDGs with organisational practices. The first lesson is to ensure that executives and managers understand what internal sustainability is, and how it is linked to good governance in terms of processes, which then has an impact on outcomes. Maintaining good governance at the level of the organisation is important for preserving public trust, which is especially necessary in the public sector, whose mandate comes from the public. Secondly, a designated unit should be created to promote sustainability practices and values within the organisation and translate this onto its outputs and outcomes. To improve the effectiveness of organisational sustainability management and practices, sustainability skills training should be offered. Thirdly, to encourage shared learning and advancement that supports Thailand's sustainable development, a platform should be created where organisations from various sectors can exchange knowledge and skills related to their practical sustainability challenges and solutions. Finally, it must be noted that an organisation's sustainability culture can only be built over time, and continuous efforts must be made to maintain it.

References

Addink, H. 2019. *Good governance: Concept and context.* Oxford: Oxford University Press.

Bank of Thailand. 2023. *Sustainable Solutions to Thailand's Household Debt Problems.* Available at https://www.bot.or.th/en/financial-innovation/sustainable-finance/household-and-sme/household-debt.html#accordion-687a5432f6-item-69b25da62b (Accessed: 10 March 2025).

Brundiers, K. and Wiek, A. 2017. 'Beyond interpersonal competence: Teaching and learning professional skills in sustainability', *Education Sciences*, 7, pp. 1–18.

Bryson, J. and George, B. 2020 'Strategic management in public administration', in Hannah, E. (ed.) *Oxford Research Encyclopedia of Politics*, Oxford: Oxford University Press, pp 1–26.

Bryson, J. M. 2018. *Strategic planning for public and nonprofit organisations: A guide to strengthening and sustaining organisational achievement.* Oxford: Wiley.

Busco, C., Riccaboni, A. and Nasi, A. 2017. 'Sustainability management beyond compliance: An empirical study on the impact of management control systems on sustainable practices', *Corporate Social Responsibility and Environmental Management*, 24(4), pp. 312–325.

da Silva, A. V. B. and Romaro, P. 2024. 'Emerging trends in sustainable management: Developing managers' skills for ESG challenges', *Revista de Gestão e Secretariado*, 15(6), pp. 1–15.

Digio 2021. Digio handbook. Available at: https://www.digio.co.th/post/digio-handbook (Accessed: 8 August 2025).

Digio, 2022. About Us. Available at: https://www.digio.co.th/about-us (Accessed: 8 August 2025).

Domingues, A. R., Lozano, R., Ceulemans, K. and Ramos, T. B. 2017. 'Sustainability reporting in public sector organisations: Exploring the relation between the

reporting process and organisational change management for sustainability', *Journal of Environmental Management*, 192, pp. 292–301.
Figueira, I. et al. 2018. 'Sustainability policies and practices in public sector organisations: The case of the Portuguese Central Public Administration', *Journal of Cleaner Production*, 202, pp. 616–630.
Galbraith, J. R. 2002. 'Organizing to deliver solutions', *Organisational Dynamics*, 31(2), pp. 194–207.
Galbraith, S. 2020. 'Galbraith, Jay R.: Master of organisation design–recognizing patterns from living, breathing organisations', in Szabla, D. (ed.) *The Palgrave handbook of organisational change thinkers*, Cham: Palgrave Macmillan, pp 631–650.
Heimler, A. 2000. 'Competition and regulation in public utilities', *OECD Economics Department Working Papers*, No. 238. Paris: OECD Publishing. Available at: https://www.oecd.org/content/dam/oecd/en/publications/reports/2000/01/privatisation-competition-and-regulation_g1ghg93e/9789264180581-en.pdf (Accessed: 10 July 2025).
Kaplan, R. S. 2005. 'How the balanced scorecard complements the McKinsey 7-S model', *Strategy & Leadership*, 33(3), pp. 41–46.
Kasikorn Bank. n.d. *Vision and Mission.* Available at: https://www.kasikornbank.com/en/about/information/pages/vision-mission.aspx (Accessed 8 August 2025).
Kasikorn Bank. n.d. *Green DNA*. Available at: https://www.kasikornbank.com/en/sustainable-development/greenDNA (Accessed: 15 December 2024).
Kasikorn Bank. 2024. *Kasikorn Sustainability Report 2023*. Available at: https://www.kasikornbank.com/th/ir/generalinformation/pages/corporate.aspx (Accessed: 24 August 2024).
Kasikorn Bank. 2025. *Kbank at a Glance*. Available at: https://www.kasikornbank.com/th/ir/generalinformation/pages/corporate.aspx (Accessed: 10 May 2025).
Kates, A. and Galbraith, J.R., 2010. *Designing your organisation: Using the STAR model to solve 5 critical design challenges*. Oxford: Wiley.
Kraft, M. E. and Furlong, S. R. 2019. *Public policy: Politics, analysis, and alternatives*. Washington, DC: CQ Press.
Mandl, U., Dierx, A. and Ilzkovitz, F. 2008. *The Effectiveness and Efficiency of Public Spending* (European Economy Economic Papers 301). European Commission.
Ministry of Foreign Affairs of Thailand. 2017a. *Sufficiency Economy Philosophy: Thailand's Path towards Sustainable Development Goals*. Available at: https://data.thailand.opendevelopmentmekong.net//dataset/sufficiency-economy-philosophy-thailands-path-towards-the-sustainable-development-goals (Accessed: 15 September 2024).
Ministry of Foreign Affairs of Thailand. 2017b. *Summary of Thailand's Voluntary National Reviews on the Implementation of the 2030 Agenda for Sustainable Development 2017*. Available at: https://data.thailand.opendevelopmentmekong.net//dataset/2560-2573 (Accessed: 16 September 2024).
Ministry of Foreign Affairs of Thailand. 2024. *Thailand's Voluntary National Review on the Implementation*. Available at https://sep4sdgs.mfa.go.th/en/ebook-category/voluntary-national-review-vnrs- (Accessed: 10 March 2025).
Mongsawad, P. 2010. 'The philosophy of the sufficiency economy: A contribution to the theory of development', *Asia Pacific Development Journal*, 17(1), pp. 123–143.

Mongsawad, P. and Thongpakde, N. 2016. 'Sufficiency economy philosophy: A holistic approach to economic development and mainstream economic thought', *Asian Social Science*, 12(7), pp. 136–142.

National Economic and Social Development Council (n.d.). *History and Role of NESDC*. [online] Available at: https://www.nesdc.go.th/en/interesting-facts/history-of-nesdc/ (Accessed 8 August 2025).

National Economic and Social Development Council n.d. *Vision/Organization Culture/NESDC Values*. [online] Available at: https://www.nesdc.go.th/en/interesting-facts/history-of-nesdc/ (Accessed 8 August 2025).

National Economic and Social Development Council, n.d. Message from the Secretary-General. Available at: www.nesdc.go.th/article_attach/article_file_20210222143231.pdf (Accessed 15 July 2025).

National Economic and Social Development Council 2019. *National Strategy 2018–2037*. Available at: https://www.nesdc.go.th/wordpress/wp-content/uploads/2025/11/PPT-AIT-.pdf (Accessed: 28 July 2024).

National Economic and Social Development Council. 2020 *History and Roles of NESDC*. Available at: https://www.nesdc.go.th/nesdb_en/ewt_news.php?nid=4258 (Accessed: 30 July 2024).

National Economic and Social Development Council, 2024. **แผนปฏิบัติราชการ สำนักงานสภาพัฒนาการเศรษฐกิจและสังคมแห่งชาติ (สศช.) ประจำปีงบประมาณ พ.ศ.** 2568 [Government Action Plan, National Economic and Social Development Council (NESDC) for the 2025 Fiscal Year]. Available at https://www.nesdc.go.th/ewt_dl_link.php?nid=16035 (Accessed: 12 December 2024).

Nation Thailand, 2025. *Rama II Road sees over 2,500 construction accidents in seven years. Nation Thailand*, 15 March. Available at: https://www.nationthailand.com/thailand/bangkok/40047456 (Accessed: 15 July 2025).

Oertwig, N., Galeitzke, M., Schmieg, H.-G., Kohl, H., Jochem, R., Orth, R. and Knothe, T. 2017. 'Integration of sustainability into the corporate Strategy', in Stark, R. Seliger, G. & Schulze, J. H. W. (eds.) *Sustainable manufacturing*, Wiesbaden: Springer Gabler, pp.175–200.

Osborne, S.P., Radnor, Z. and Nasi, G. 2013. 'A new theory for public service management? Toward a (Public) service-dominant approach', *The American Review of Public Administration*, 43(2), pp. 135–158.

Pattaya Mail 2022. 'Bangkok reports 30% rise in number of homeless persons', *Pattaya Mail*, 13 October.

Royal Thai Embassy, Permanent Mission of Thailand to the United Nations, New York 2024. *Sufficient Economy Philosophy for SDGs*. Available at: https://unmissionnewyork.thaiembassy.org/en/content/38134-sep-for-sdgs (Accessed: 23 June 2025).

S&P Global. 2024. *The Sustainability Yearbook 2024: Rankings*. S&P Global ESG. Available at: https://www.spglobal.com/esg/csa/yearbook/2024/ranking/ (Accessed: 15 July 2025).

Scott, W. and Davis, F. 2007. *Organizations and organizing: Rational, natural and open system perspectives*. Hoboken, NJ: Pearson Prentice-Hall.

Stock Exchange of Thailand, 2023. *Criteria and Guideline: Sustainability Assessment 2023 (SET ESG Ratings)*. Bangkok: Stock Exchange of Thailand. Available at: https://setsustainability.com//download/b3zk5xnjpwuh8rt (Accessed: 11 May 2024).

Suttipun, M. 2018. 'The influence of corporate governance, and sufficiency economy philosophy disclosure on corporate financial performance: Evidence from Thailand', *Asia-Pacific Journal of Business Administration*, 10(1), pp. 79–99.

The Government Public Relations Department, 2016 *Thailand's 20-Year National Strategy and Thailand 4.0 Policy*. Available at: https://data.thailand.opendevelopmentmekong.net//dataset/thailand-s-20-year-national-strategy-and-thailand-4-0-policy (Accessed: 15 September 2024).

United Nations (2015). *Transforming our world: The 2030 Agenda for Sustainable Development*. Retrieved from https://sdgs.un.org/2030agenda (Accessed: 15 September 2024).

United Nations Economic and Social Commission for Asia and the Pacific (UNESCAP). (2009). *What Is Good Governance*. Available at: https://www.unescap.org/sites/default/files/good-governance.pdf (Accessed: 14 July 2025).

7

GOOD GOVERNANCE IN SUPRANATIONAL SETTINGS

The principle of participation and responsiveness to public demands in the EU

Anastasia Ershova

Introduction

The European Union (EU) is the most complex supranational integration project in the world. In practice, its governance competencies vary from non-existent in policy areas such as education, to exclusively supranational in the domain of market competition. Tracing its origins to the 1950 Schuman's Declaration and the subsequently established European Coal and Steel Community, the Union represents a gradually evolving supranational governance set up which offers valuable insights and is used as a benchmark by other regional blocs outside the European continent. Although there are a growing number of studies underlining how the EU is oftentimes used as an implicit standard which other blocs and integration projects should adopt or emulate (see e.g. Börzel and Risse, 2009; Jetschke and Murray, 2012), it is surprisingly rare that the practices and principles of governance in the EU itself are benchmarked against broader standards of good governance. This chapter takes the first step in locating the discussion of EU governance within the context of good governance criteria. It primarily zooms in on the role of participation as a precondition for good governance and highlights possible limitations in the current status quo of the Union in this regard.

The EU governance structure has been criticised for being 'too distant' and allegedly adopting 'policies that are not supported by a majority of citizens in many or even most Member States' (Follesdal and Hix, 2006, 537). This has led to a compelling image of the Union as a polity suffering from a pervasive democratic deficit. Despite the increased powers of the European Parliament (EP) – the only institution of the EU where citizens hold electoral control

DOI: 10.4324/9781003467694-9

over its members – the other key institutional players (the Council of Ministers and the European Commission) are often seen as being insulated from public scrutiny, because of their appointment procedures and the complexity of the institutional set-up. The Council of Ministers has indeed long been creating EU policies in isolation from public pressures, as the EU populace has had only limited opportunities to observe and assess Ministers' actions in the Council during the policymaking process (Bailer et al., 2015). Similarly, the Commission has been portrayed as a technocratic, non-majoritarian institution (e.g. Egeberg et al., 2014; Majone, 2005; Radaelli, 2017), hardly concerned with public pressures as neither its term nor the appointment of its members are subject to direct public accountability mechanisms.

Such a narrative of the EU governance system clashes with the European Commission's own 'good governance' criteria set out as guiding principles for the reform of EU governance towards greater legitimacy. Despite the ever-changing scope of 'good governance' as a concept, the EU (European Commission, 2001) has followed other international institutions such as the OECD (2006) in listing participation, transparency, and effectiveness as core good governance criteria. However, while effectiveness or transparency can be more readily applied to supranational systems, the *de facto* application of the participation principle to policymaking in the complex system of the EU is often unclear. This, in turn, raises a number of questions about the willingness and capacity of the Union to adhere to its own principles of good governance.

The primary focus of this chapter is the EU executive – the European Commission – and the extent to which this institution maintains legitimacy through its input responsiveness to public demands. In line with the extant literature, this chapter views responsiveness as a dynamic process where political and bureaucratic actors actively aim to address public demands and concerns. It captures the ability and willingness of policymakers to react to and reflect changes in public opinion in their policymaking behaviour (see e.g. Soroka and Wlezien, 2010). Specifically, input responsiveness in the EU context can be thought of as 'actions taken by relevant actors during the EU decision-making process that respond to public preferences' (Meijers et al., 2019, p. 1726). The responsiveness to public preferences, shaped by political debate and competition between institutions, adds to the ability of institutional actors to maintain their legitimacy, as it secures public acceptance (see Schmidt, 2022, 25–27). The link between input responsiveness and legitimacy is particularly salient for non-majoritarian institutions insulated from electoral controls, which the public did not directly empower with decision-making authority (see Thatcher and Stone-Sweet, 2002).

Sometimes characterised as a 'runaway bureaucracy,' the Commission possesses a unique power of legislative initiative, shaping the agenda of the Union and putting forward measures that can translate into policies implemented across the member states. Despite a well-established literature that aims to understand the behaviour and choices of this institution, there is a surprising lack of studies examining the extent to which general principles of good governance shape the Commission's policymaking. This is where this chapter aims to make its key contribution. First, it engages with the concept of good governance, its adoption by the EU, and the specific interpretation the EU applies to it. Thereafter, it links the principles of good governance to the ongoing research on institutional responsiveness. This chapter proceeds with a review of recent contributions to the literature on EU governance which evidence patterns of responsiveness to public demands within the European Parliament and Council of Ministers. This chapter then turns its attention to the EU executive branch and engages with public opinion indicators from Eurobarometer surveys and with recent findings in the literature which suggest that the supranational bureaucracy tends to maintain responsiveness in its behaviour, despite expectations to the contrary (e.g. Zhelyazkova et al., 2019). Finally, this chapter highlights the limitations of existing studies, underlining how cross-national variation in levels of public support for and salience of specific policy domains limits the applicability of good governance criteria to the behaviour of the EU agenda-setter.

Good governance, legitimacy, and responsiveness in the EU

Since the introduction of the term by the World Bank in 1989, 'good governance' has been actively promoted by local, national, international, supranational, and global organisations across different policy domains. Initially formulated as a set of criteria for countries to qualify for aid, 'good governance' was used to describe accountability in administration, efficiency in public services, technocracy, a dependable judicial system, and an overall balance between government and the governed (e.g. Williams and Young, 1994; Van Doeveren, 2011). As the term spread across international organisations, its initial definitions have been continually adapted. For instance, the International Monetary Fund (IMF) stresses the importance of tackling corruption, describing it as 'an essential element of a framework within which economies can prosper' (International Monetary Fund, 1998, p. V). The UN definition underscores the importance of consensus-oriented participatory governance (UNESCAP, 2009), while the OECD (2006) highlights the role of the rule of law, democracy, and representation (Agere, 2000). Similarly, the EU has referred to good governance in its Copenhagen accession criteria, which emphasise the importance of democracy, respect for minorities, and human rights.

While the scope of the concept appears to be ever-changing, there are clearly core values that remain shared across contexts and actors: transparency, accountability, effectiveness, and the rule of law. The rule of law principle facilitates a stable environment and development by guaranteeing human rights and upholding property rights. Transparency and accountability contribute to effective decision-making and facilitate the responsiveness of the political system. This, in turn, furthers the legitimacy of the governing system, and by extension, its policies.

The interrelationship between these criteria is clearer in the context of a democratic state, where the distance between governing institutions, civil society, and the public is small, and the roles of different institutions within the governing system are clearly defined. In such a context, good governance could even be thought of as a 'public administration process that maximizes public interest,' aimed at bringing together the State, the citizens, and their relationship with civil organisations through the collaborative management of public life (see e.g. Keping, 2018). In other words, good governance hinges on political actors' participation in public life. Participation ensures the legitimacy of actions taken as a result of citizen involvement.

However, the extent to which transparency, accountability, and effectiveness may foster responsiveness and legitimacy in a multilevel governing system is less straightforward. In the case of the EU, its governance system is highly complex, characterised by fuzzy responsibility chains (Ershova, 2025), a blurry distribution of power between EU institutions, and a fragmentation of competencies between national and supranational elites. In such an environment, institutions may lack the incentives to maintain responsiveness to citizens and/or civil society, which in turn translates into undermined input legitimacy (see e.g. Schmidt, 2022) of the governance system as a whole. 'Input' legitimacy is one of the sources from which EU institutions may draw (Höreth, 2001). It differs from output legitimacy, which refers to the outcomes of governance, or the extent to which it solves existing problems. By contrast, 'input legitimacy' is more about governance processes, including citizens' participation and transparency.

The extent to which input legitimacy matters for good governance was not always explicitly recognised by EU institutions. However, by the late 1990s to early 2000s, the Union had faced a series of political crises and unforeseen challenges. Diminished public confidence following the resignation of the Santer Commission in 1999 (Ringe, 2005) and the resulting low turnout at European elections signalled growing public disillusionment with the supranational institutions. Followed by the unexpected 'No' vote of the Irish electorate on the ratification of the Nice Treaty, these events indicated to the EU 'a need for urgent action to adapt governance under the existing treaties, but also for a broader debate on the future of Europe [...]' (European Commission, 2001). Having completed extensive consultations with regional and local actors, civil society, citizens, national parliaments, and member states' governments (for details see Sloat, 2003), the European Commission put forward the White Paper on Governance in July 2001. It laid down five principles of good governance in the EU: openness, participation, accountability, effectiveness, and coherence (European Commission, 2001).

While aiming to address the EU's governance crisis through a substantive reform of EU policymaking, the 2001 White Paper on Governance emphasised that 'policies should no longer be decided at the top. The legitimacy of the EU now lies with the participation of its citizens' (European Commission, 2001). The White Paper assumed that increased civil society participation and consultations, coupled with transparency and public access to the documents produced by core EU institutions (for details, see Regulation (EC) 1049/2001), would facilitate democratic processes in the EU. The strong emphasis on the role of citizens and civil society is hardly surprising. While in its early years, the Union enjoyed what is called 'permissive consensus,' the institutional and political crises of the early 2000s signalled a growing dissensus among the public regarding further EU integration. What scholars describe as the awakening of the 'sleeping giant' of public opinion scepticism (Van Der Eijk and Franklin, 2004; deVries, 2007) has shaped a new reality for EU policymaking. In this new environment, output legitimacy stemming from the efficiency of supranational policies has become insufficient to sustain the legitimacy of the governance system overall. Engagement with citizens' views and mobilised interests at different policy stages is essential to increase the input legitimacy of EU policymaking. This, in turn, is key to the survival of governing institutions, safeguarding their ability to effectively deliver and implement policies, as well as securing adherence to and compliance with them (Dellmuth and Tallberg, 2015; Tallberg and Zürn, 2019; Buchanan and Keohane, 2006). To put it differently, the *participation* of citizens and stakeholders enables governing institutions to build, demonstrate and maintain their responsiveness to the articulated demands, hence fostering *input legitimacy of* the produced measures and adding to the legitimacy of the system overall.

However, how can one understand whether the principles of good governance and specifically engagement with the public and civil society have translated into substantive changes in policies or institutional behaviour? In his prominent contribution focused on issues of legitimacy, Scharpf (1999) maintains that input legitimation of the EU could be judged in terms of the EU's responsiveness to citizens' and stakeholders' demands.[1] Similarly, Pennock (1952, 790) defined legitimacy as 'reflecting and giving expression to the will of the people.' This link between participation, responsiveness and input legitimacy has inspired scholars to scrutinise whether (and if so, to what extent) the behaviour and policy choices of EU institutions are shaped by public pressure.

Responsiveness in the legislative branch of the EU: the European Parliament and the Council of Ministers

The EU system of governance is complex, with overlapping institutional responsibilities. Nevertheless, one can still clearly differentiate between the main branches of governance. The Court of Justice of the European Union (CJEU) represents the judicial branch, the EP and the Council function as the chambers of the legislative branch, while the Commission acts as the EU executive. The CJEU interprets EU law, ensures its consistent application across the Union, and settles disputes between the EU and its member states. While it is a key EU institution, it plays only a limited role in routine policymaking. In contrast, the EP and the Council hold the power to amend, adopt, or reject policy proposals initiated by the European Commission.[2]

There is a growing body of literature focusing on the legislative branch of the EU and its responsiveness. Examining the behaviour of the European Parliament, Lo (2013) highlights how Members of the European Parliament (MEPs) tend to adjust their voting patterns to align with demands from domestic audiences. For example, the study shows that after the Nice Treaty referendum in Ireland, Irish MEPs responded to their constituents' critical attitudes towards the EU by voting more conservatively in parliamentary roll-call votes. Similarly, Ministers in the Council have become more constrained by public preferences and can no longer rely on citizens' indifference to their positions and decisions during the legislative process.

Existing research demonstrates that both institutions take citizens' attitudes into account at various stages of the decision-making process. These attitudes include views on European integration in general (Hagemann et al., 2017; Wratil, 2018) as well as on specific policies (Hobolt and Wratil, 2020; Mühlböck and Tosun, 2018; Schneider, 2019a, 2019b). For instance, drawing on expert survey estimates from the Decision-making in the European Union (DEU) dataset (Thomson et al., 2012), Wratil (2018) demonstrates that governments' negotiation positions within the Council are responsive to domestic public opinion on left-right issues in the run-up to national elections and on issues related to European integration when these become domestically salient. Similarly, relying on a novel approach that examines the Council through video footage of its public deliberations (DICEU – Debates in the Council of the European Union), Hobolt and Wratil (2020) show evidence that public opinion shapes ministerial policy positions, particularly on domestically salient issues. Furthermore, analyses of Council debates suggest that domestic public views influence the willingness of pro-EU governments to compromise during the Council's legislative deliberations (Mariano and Schneider, 2022; Wratil et al., 2023).

Taken together, these studies provide an optimistic portrayal of the EU's legislative branch, represented by the EP and the Council, as consistently maintaining responsiveness to public demands. This responsiveness enhances the input legitimacy of the Union and aligns with the principles of good governance, particularly participation. However, these findings also underscore the strong connection between domestic audiences and the behaviour of representatives in supranational institutions. Factors such as the electoral accountability imposed on Ministers in the Council (Kleine and Minaudier, 2017; Hagemann et al., 2017) and the influence of domestic party politics on MEPs' actions (e.g. Hix, 2004; Mühlböck and Yordanova, 2017; Costello and Thomson, 2016) are essential in ensuring the responsiveness of institutional actors.

Responsiveness in the EU executive branch: the European Commission

The executive arm of the EU governance system is represented by the European Commission. Unlike traditional executive branches within state systems, the Commission possesses unique monopoly power to propose and shape the content of any new legislation granted to it by the EU Treaties (European Union, 1992, Art. 17(2)). Unlike the EP and the Council, the Commission is not subject to electoral pressures at either the EU or domestic levels and is expected to act in the interest of the Union as a whole. This unique institutional role defines the Commission's distinctive agenda-setting power within the EU's institutional constellation.

During the early years of European integration, the Commission was perceived as the embodiment of technocracy – driven by expertise, lacking electoral accountability, and free of political allegiances. It was seen as deriving legitimacy not from responsiveness to public demands but from the effectiveness and expert knowledge characterising its outputs (Schmidt, 2022). In recent years, however, the EU agenda-setter has become a self-proclaimed political actor (Kassim and Laffan, 2019; Koop et al., 2022), spurring a growing body of empirical research investigating the conditions that shape the Commission's behaviour.

Although consistently pro-integrationist in its preferences (König and Pöter, 2001; Rauh, 2019), the Commission's actions appear increasingly influenced by public input. Evidence suggests that its activities at the agenda-setting stage of EU policymaking correlate with citizens' preferences and the salience the public assigns to EU policies (Koop et al., 2022; Rauh, 2016, 2019). For instance, analysing the Commission's policy initiatives on consumer protection, Rauh (2019) shows that EU politicisation[3] and policy salience contribute to more consumer-friendly proposals (see also Rauh, 2016). Additionally, Reh et al. (2020) demonstrate that the Commission's attentiveness to public opinion can extend beyond the legislative proposal stage, as reflected in decisions to withdraw proposals when they spark public discontent. Similarly, Koop et al. (2022) find that the Commission prioritises publicly salient policy issues in its political programmes and annual working plans. Notably, legislation that expands EU authority is more likely to feature among these priorities, driven by a need to signal responsiveness to broader audiences beyond experts and international politicians.

In a more recent study, Yordanova et al. (2024) examine how the Commission's proposals to expand EU authority respond to public demands under growing politicisation pressures. They define the expansion of EU authority as 'an increase in either the scope of its governance competences by establishing EU legislation or programs in previously unaffected areas, or the level of its existing governance competences within a policy domain relative to national authority' (Hagemann et al., 2017, 855). To understand whether there is a link between changes in the proposed EU authority and public demands, Yordanova et al. (2024) compiled a novel dataset. They turned to the legislative proposals put forward by the European Commission and set to be scrutinised by both the EP and the Council as co-legislators of equal footing, under the so-called Ordinary Legislative Procedure (OLP). The authors focused on the files proposed after the latest Treaty reforming the governance structure of the EU, ratified in 2009, commonly known as the Lisbon Treaty. Drawing on these data, the study showed that the Commission's ambition to expand EU authority aligns with rising public demands for EU action in specific policy areas. However, this responsiveness diminishes when public polarisation and the mobilisation of diverse interest groups increase, a consequence of the heightened politicisation of European politics. These hindering effects emerge because divided public opinion and competing demands limit the Commission's ability to craft policy proposals satisfying different groups. The findings further show that increased salience across the EU facilitates Commission responsiveness. High-salience policy areas necessitate action, where the Commission, with its legislative initiative and pro-integration preferences, is well-positioned to pursue deeper integration. Such actions may not always be perceived as responsive but could be deemed 'responsible,' aimed at addressing pressing policy problems.

Yordanova et al.'s (2024) study results are encouraging – even an institution with limited public accountability and no exposure to domestic politics strives to maintain input responsiveness, thus contributing to the EU's input legitimacy. One might argue that the Commission's drive to support integration in response to urgency, salience, and public demands reflects functional pressures. As the EU matures, policy interdependencies create incentives for the supranational bureaucracy to propose solutions attuned to emerging challenges. By addressing these policy needs, the Commission acts as an apolitical, benevolent planner focused on effective governance. However, this behaviour can also be interpreted as strategic and self-preserving, especially when evaluated in relation to the expansion of EU authority.[4] The Commission's power depends on the EU's continued integration and prosperity (Franchino, 2007; Rauh, 2021; Ershova, 2025). As the Union's authority deepens or broadens, so does the Commission's agenda-setting influence across policy domains. However, this growing capacity may bring the additional burden of securing legitimacy, especially for policies addressing urgent and salient problems. Whether the Commission's responsible, expert-driven solutions under conditions of high salience can counterbalance a lack of input responsiveness remains an open question.

Challenges to responsiveness in the European Commission

Most studies examine Commission responsiveness using aggregated indicators of issue and/or policy salience. While this approach captures average trends across member states, it may obscure significant variations within states. For example, drawing on Eurobarometer data (2009 –2019) compiled by Yordanova et al. (2024), differences in policy issue salience between member states become apparent. Focusing on policy domains defined in the European Union Policy Agendas Project (Alexandrova et al., 2014), Figure 7.1 illustrates that even when averaged over time, public salience attached to EU action varies significantly across member states. In states where a policy issue is highly salient, Commission initiatives may resonate with citizens, fostering the legitimacy of proposed solutions. Conversely, in states where the issue is of low concern, the same policy may fail to appeal to the public, or even worse, trigger public rejection of the EU-level initiative, undermining the legitimacy of the policy overall. In high-salience states, the Commission may bolster output legitimacy by avoiding the costs of inaction and demonstrating effective policymaking (Blumenau and Lauderdale, 2018). However, in low-salience states, claims of output legitimacy may be less justifiable.

This tension challenges the Commission's commitment to its own principles of good governance. A core tenet of good governance posits that policymaking should be effective. The Commission defines effectiveness as ensuring that 'decisions need to be taken at the appropriate level and time, and deliver what is needed' (European Commission, 2001). Legislative proposals extending EU authority in areas of high average salience may embody this principle. However, recognizing variation in salience across member states raises questions about the principle's consistent application. Divergent public needs across the Union complicate the Commission's ability to simultaneously address varying levels of urgency, challenging its adherence to the principle of effectiveness in governance.

Furthermore, while studies thus far suggest that the EU executive is receptive to public input (Yordanova et al., 2024; Rauh, 2019), helping it build the

Figure 7.1 Issue salience across EU member states, based on Eurobarometer responses averaged across 2009–2019. Author's own, based on data from Yordanova et al.

Note: Salience is measured as the inverse proportion of 'Don't Know' responses over the total number of responses to Eurobarometer questions targeting specific policy areas. This is based on the assumption that when respondents do not have an opinion on a given issue, they regard it as non-salient (see Yordanova et al., 2024, for discussion). The policy area of 'Culture' was excluded due to a lack of observations.

image of a responsive agenda-setter, the evidence of this responsiveness is largely based on analyses that utilise aggregated estimates of public demands averaged across all member states. By highlighting the variance in public support across individual states (see Figure 7.2), we can flesh out the differences in perceived desirability of EU action in different member states and policy areas. These differences do not necessarily undermine the findings that the Commission has been striving to maintain responsiveness and adhere to the principle of participation, which seeks to account for the voices of the European populace. However, they raise an important question: which voices, and why, have proven decisive when policy proposals were formulated and their level of ambition adjusted? What are the conditions under which the demands of citizens in a particular member state become pivotal in shaping EU-wide policy?

There is only limited research tackling these challenging and puzzling questions. For instance, Wratil (2018) shows that there is a certain degree of congruence between EU policies and state-level public opinion, which becomes more pronounced when the issue salience and public consensus increase. However, this static congruence (for discussion see Yordanova et al., 2024; Lax and Phillips, 2012; Louwerse and Andeweg, 2020) only captures whether policies correlate with public preferences, but not whether shifts in public attitudes are followed by changes in policies. Nor do they explain what happens with policy content when issue salience in several member states reaches high(-er) levels, but diverging public opinion across member states demands contradictory substantive actions to be taken at the EU level. For instance, based on Figure 7.1, the policy area of 'Labour' is generally seen as of high salience across EU states, with the stark exception of Poland. However, as Figure 7.2 shows, the support of the public for EU action in this area drastically differs between, for example, Southern and Northern European states. Hence, participation and input from which citizens will shape the policy proposed by the multilevel governance system of the EU and the non-elected Commission?

Drawing on the literature on EU legitimacy, one could hypothesise that if responsiveness to certain public demands comes at the expense of responsiveness to others, policy efficiency might compensate for the lack of citizen participation (e.g. Schmidt and Wood, 2019). However, this compensation mechanism assumes that the resulting policy and its outcomes would sufficiently meet the expectations of the public whose input was disregarded during the formulation stage. Although this assumption may hold under certain conditions, it is relatively strong, as it treats public preferences regarding policy as stable, easily observable, and implies equal importance is attached to both the input and output of governance systems. Therefore, only by relaxing these assumptions and conducting a rigorous empirical investigation can one confidently argue that the EU Commission's input responsiveness to only some public views does not undermine policy legitimacy, provided that the unaddressed public demands are satisfied through demonstrably effective policy outcomes. In fact, claims of professionalism, unbiasedness, and the high-level skills used in the production of policy measures are often

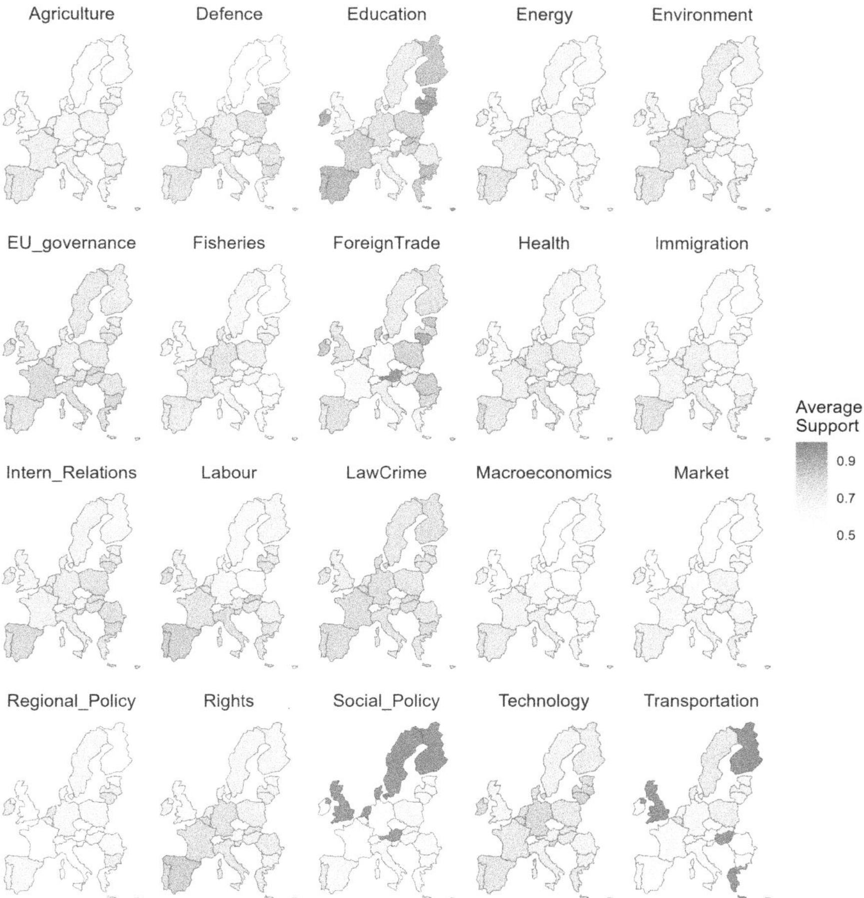

Figure 7.2 Public support for EU action averaged across 2009–2019. Author's own, based on data from Yordanova et al. (2024).

Note: Public Support is measured as the proportion of responses indicating support for EU action. The Eurobarometer post-stratification weights were applied. The policy area of 'Culture' was excluded due to a lack of observations.

used within bureaucracies (national and supranational) to justify their policy choices. Such claims defend the power of non-elected officials and policymakers (Schmidt, 2020; Thatcher and Stone-Sweet, 2002) which may well choose a policy option at odds with expressed public demands. Under such circumstances, participation as a characteristic of good governance becomes moot, as the EU Commission exploits output legitimacy that rests on its own professionalism and high skill instead of responding to the inputs of the public.

Experts in EU studies may argue that the extent to which the Commission could disregard the preferences of some parts of the public at the expense of others is in fact constrained due to existing institutional arrangements in the EU and the power of electoral controls that are indirectly applied to the Commission. Specifically, while being insulated from direct electoral pressure, the Commission is appointed and approved by the directly elected EP and the Council of Ministers. While the members of the latter are chosen via domestic electoral dynamics, the EP is directly elected by the European public. The EP's role as a co-legislator that co-adopts legislative proposals and can amend, accept, or refuse proposals enables the EP to act as a guardian of input responsiveness across EU policymaking and secure policy outcomes that account for varying demands of the EU publics.

However, the representativeness of elected institutions is directly dependent on the participation of the public being represented. For the EP to act as a champion of EU citizens, whose will is ideally uniformly reflected in the produced policies, it needs direct engagement from those very citizens at the time of election. Based on turnout indicators at the EU elections presented in Figure 7.3, participation is hardly uniform or stable across time or across countries. These patterns do raise questions about the extent to which the EP's actions are geared to address the interests of the broader EU public. These patterns also add to concerns regarding whose voices prevail, and what mechanisms could be used to secure the interests of politically disengaged citizens, but without undermining the governance system that relies on actors' participation as a requirement of good governance.

To sum up, there is a growing body of literature analysing the responsiveness of the European Commission at various stages of policymaking. The existing findings are encouraging and challenge the typical portrayal of the Commission as a detached and non-responsive technocracy. Evidence suggests that the Commission listens to public demands when shaping its proposals and calibrating the ambitions of expanding EU authority. The Commission is also susceptible to the pressures of the politicisation of European politics. However, critical gaps remain in our understanding, preventing us from conclusively asserting that the principle of participation, as a good governance criterion, is consistently upheld by this institution. Furthermore, the inconsistent and unstable turnout rates in European elections signal that the issue of participation may have further consequences for governance in the EU. Low turnout limits the extent to which the directly elected Parliament can act as a representative of the European public in the true sense of the word.

Concluding remarks

Principles of good governance have long captured the attention of academics, political experts, and global entities. These principles have been modified to

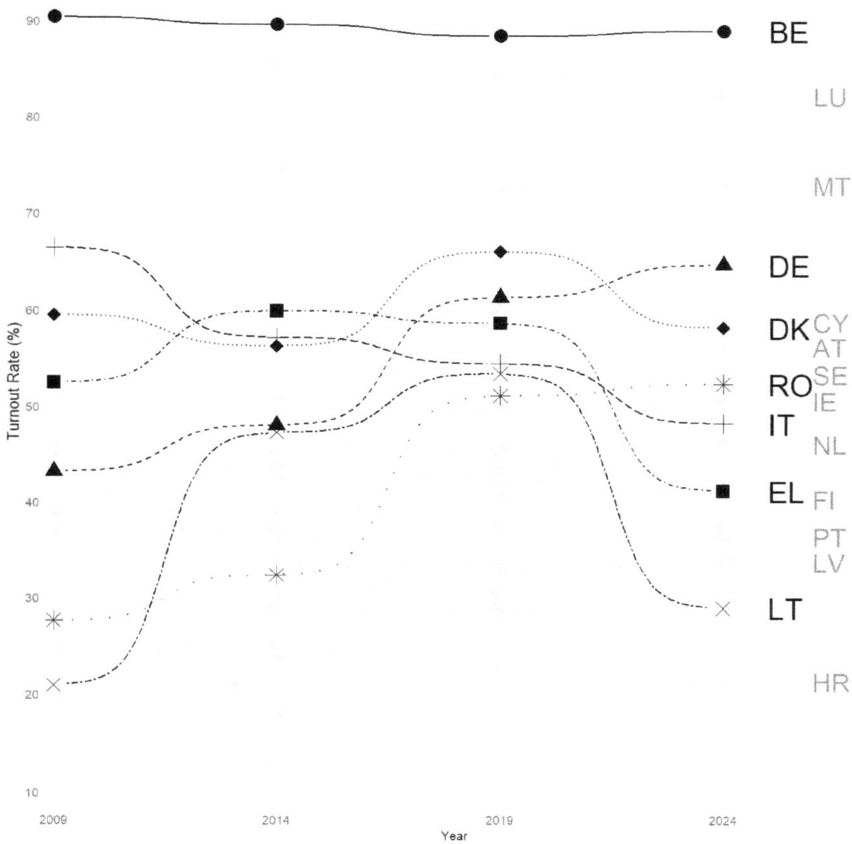

Figure 7.3 Voter turnout in percentage points across EU 27. Author's own, based on data from the European Parliament at https://results.elections.europa.eu/en/. Highlighted countries represent different regions across the EU.

suit different contexts and have served as benchmarks for governance reform within the EU. Researchers examining development policies have considered whether, and to what extent, EU policies in development and external affairs integrate and adhere to good governance standards (e.g. Börzel et al., 2008).

Surprisingly, the efforts of EU institutions to foster the legitimacy of EU governance have rarely been examined as a means of adhering to good governance criteria. This chapter took the first step towards developing such research. First, it outlined the differing definitions of good governance across various international institutions and examined which good governance principles have been adopted by the EU, and how. It drew on the link between the principles of participation and effectiveness, as encapsulated in the

Commission's White Paper, and the rich literature on policy responsiveness to highlight the extent to which EU political actors appear to follow the criteria of good governance. Thereafter, it examined existing findings on how public opinion and a politicised environment have shaped the responsiveness of the EP, the Council, and the Commission. This chapter has shown that there is growing evidence that both the legislative and executive branches of the EU consider public views at different stages of the policymaking process. Overall, this is a highly encouraging message, as it challenges the longstanding narrative that the EU governance system suffers from a democratic deficit (Follesdal and Hix, 2006).

However, as this chapter zoomed in on recent findings focused on the responsiveness of the EU Commission, it argued that one should move beyond aggregated analyses of cross-EU trends in public opinion and proxies for politicisation. Using Eurobarometer data, it highlighted the differences (averaged over time) in the salience of policy areas and the public's support for EU policy action across EU member states. Stark contrasts in both the urgency of action, and public demands for action across different European countries raised important questions about the aggregated patterns of responsiveness emphasised in existing studies. Are the patterns of responsiveness identified in recent literature driven by specific member states? What are the conditions under which this happens, and what are the implications for EU policy legitimacy? Extant studies have not yet addressed these questions. In fact, the answers appear to be even less clear when the Commissions' role as the EU executive and legislative agenda-setter is examined in the context of institutional constraints imposed by the EP, its own legitimacy affected by the varying rates of participation in EU elections across countries and through time.

The changing public views in the EU underscore the importance of understanding the extent to which the EU meets its own criteria of good governance, devised over 20 years ago in response to growing disillusionment with the Union and a series of crises. Since 2001, the EU has systematically aimed to create a better image of transparency and efficiency. The adoption of transparency regulations (e.g. Regulation (EC) 1049/2001), the institutionalisation of stakeholder consultation as an essential step in policymaking, and the growing role of citizens' initiative, among others, certainly signal substantial efforts on the Union's part to foster openness, accountability, and participation – the key criteria highlighted by the Commission in its own definition of 'good governance.' However, while EU initiatives and reforms may be geared towards these principles, the extent to which the EU agenda-setter itself can claim to adhere to the aforementioned principles remains, at best, opaque. Only by systematically addressing these questions can one reasonably argue that the criteria of good governance put forward in the European Commission's White Paper have been realised in its legislative activities. Advancing research on these issues is particularly crucial given the recent crises the EU has faced and the growing pressure of Eurosceptic attitudes across member states.

Notes

1 This view aligns well with the work of David Easton (1957), who defined input into the 'black box' of the political system in terms of citizens' demands and support, which are conveyed through elections, as well as through citizen identity and a sense of system legitimacy.
2 Depending on the procedure, the role and powers of the EP vary. However, under the most common procedural rule – the Ordinary Legislative Procedure (OLP), formerly known as co-decision – the EP and Council act as co-legislators.
3 Here, I follow De Wilde and Zurn (2012, p. 139), who define politicisation, in general term as 'the demand for or the act of transporting an issue into the field of politics – making previously apolitical matters political.'
4 Following Yordanova et al. (2024), EU authority expansion is measured along a spectrum: maintaining the current level of EU authority (Status Quo = 0) to expanding EU authority (Expansion = 1).

References

Agere, S. (2000), *Promoting good Governance: Principles, Practices and Perspectives*, London: Commonwealth Secretariat.

Alexandrova, P., Carammia, M., Princen, S. and Timmermans, A. (2014), 'Measuring the European council agenda: Introducing a new approach and dataset', *European Union Politics* 15(1), 152–167.

Bailer, S., Mattila, M. and Schneider, G. (2015), 'Money makes the EU go round: The objective foundations of conflict in the Council of Ministers', *JCMS: Journal of Common Market Studies* 53(3), 437–456.

Blumenau, J. and Lauderdale, B. E. (2018), 'Never let a good crisis go to waste: Agenda setting and legislative voting in response to the EU crisis', *The Journal of Politics* 80(2), 462–478.

Börzel, T. A., Pamuk, Y. and Stahn, A. (2008), 'Good governance in the European Union', Berlin Working Paper on European Integration No. 7. https://refubium.fu-berlin.de/bitstream/handle/fub188/19026/2008-7boerzelpamukstahn.pdf;jsessionid=045f1b50be949ff7516412f547876c41?sequence=1

Börzel, T. A. and Risse, T. (2009), Diffusing (inter-) regionalism: The EU as a model of regional integration. (KFG Working Paper Series, 7). Berlin: Freie Universität Berlin, FB Politik- und Sozialwissenschaften, Otto-Suhr-Institut für Politikwissenschaft Kolleg-Forschergruppe "The Transformative Power of Europe". https://nbn-resolving.org/ urn:nbn:de:0168-ssoar-368027

Buchanan, A. and Keohane, R. O. (2006), 'The legitimacy of global governance institutions', *Ethics & International Affairs* 20(4), 405–437.

Costello, R. and Thomson, R. (2016), 'Bicameralism, nationality and party cohesion in the European Parliament', *Party Politics* 22(6), 773–783.

de Vries, C. E. (2007), 'Sleeping giant: Fact or fairytale?: How European integration affects national elections', *European Union Politics* 8(3), 363–385.

De Wilde, P. and Zürn, M. (2012), 'Can the politicization of European integration be reversed?'* *JCMS: Journal of Common Market Studies* 50, 137–153.

Dellmuth, L. M. and Tallberg, J. (2015), 'The social legitimacy of international organisations: Interest representation, institutional performance, and confidence extrapolation in the united nations', *Review of International Studies* 41(3), 451–475.

Easton, D. (1957), 'The political system', *World Politics* 9(3), 383–400.

Egeberg, M., Gornitzka, Å. and Trondal, J. (2014), 'A not so technocratic executive? everyday interaction between the European parliament and the Commission', *West European Politics* 37(1), 1–18.

Ershova, A. (2025), More than one agent? Authority expansion and delegation dynamics in the EU. *Regulation & Governance*. Online First. https://doi.org/10.1111/rego.12652

European Commission (2001), 'Communication from the commission of 25 July 2001 "European governance – a White Paper"'. https://eur-lex.europa.eu/EN/legal-content/summary/white-paper-on-governance.html

European Union (1992), 'Consolidated version of the treaty on European Union'. Available at: https://eur-lex.europa.eu/legal-content/EN/ALL/?uri=CELEX:12012M/TXT. Last accessed 18 March 2021.

Follesdal, A. and Hix, S. (2006), 'Why there is a democratic deficit in the EU: A response to Majone and Moravcsik', *JCMS: Journal of Common Market Studies* 44(3), 533–562.

Franchino, F. (2007), *The powers of the Union: delegation in the EU*, New York, Cambridge: Cambridge University Press.

Hagemann, S., Hobolt, S. B. and Wratil, C. (2017), 'Government responsiveness in the European Union: Evidence from council voting', *Comparative Political Studies* 50(6), 850–876.

Hix, S. (2004), 'Electoral institutions and legislative behavior: Explaining voting defection in the European parliament', *World Politics* 56(02), 194–223. https://www.journals.cambridge.org/abstractS 0043887100004275

Hobolt, S. B. and Wratil, C. (2020), 'Contestation and responsiveness in EU Council deliberations', *Journal of European Public Policy* 27(3), 362–381.

Höreth, M. (2001), 'The European Commission's White Paper on governance: A "tool-kit" for closing the legitimacy gap of EU policy-making? discussion paper', Zentrum für Europäische Integrationsforschung Center for European Integration Studies. https://d-nb.info/1263758851/34

International Monetary Fund (1998), *Good Governance: The IMF's Role, Pamphlet Series*, Washington, DC: International Monetary Fund.

Jetschke, A. and Murray, P. (2012), 'Diffusing regional integration: The EU and Southeast Asia', *West European Politics* 35, 174–191.

Kassim, H. and Laffan, B. (2019), 'The Juncker presidency: The political commission in practice', *Journal of Common Market Studies* 57, 49.

Keping, Y. (2018), 'Governance and good governance: A new framework for political analysis', *Fudan Journal of the Humanities and Social Sciences* 11, 1–8.

Kleine, M. and Minaudier, C. (2017), 'Negotiating under political uncertainty: National elections and the dynamics of international Co-operation', *British Journal of Political Science* 49(1), 315–337.

König, T. and Pöter, M. (2001), 'Examining the EU legislative process: The relative importance of Agenda and Veto Power', *European Union Politics* 2(3), 329–351.

Koop, C., Reh, C. and Bressanelli, E. (2022), 'Agenda-setting under pressure: Does domestic politics influence the European Commission?' *European Journal of Political Research* 61(1), 46–66.

Lax, J. R. and Phillips, J. H. (2012), The democratic deficit in the states. *American Journal of Political Science* 56(1), 148–166.

Lo, J. (2013), 'An electoral connection in European parliament voting', *Legislative Studies Quarterly* 38(4), 439–460.

Louwerse, T. and Andeweg, R. B. (2020), 'Measuring representation: Policy congruence', In M. Cotta and F. Russo (Eds.), *Research Handbook on Political Representation*. Cheltenham: Edward Elgar Publishing, pp. 276–288.

Majone, G. (2005), *Dilemmas of European Integration: The Ambiguities and Pitfalls of Integration by Stealth*, Oxford: OUP.

Mariano, N. and Schneider, C. J. (2022), 'Euroscepticism and bargaining success in the European Union', *Journal of European Public Policy* 29(1), 61–77.

Meijers, M. J., Schneider, C. J., & Zhelyazkova, A. (2019). Dimensions of input responsiveness in the EU: actors, publics, venues. Journal of European Public Policy, 26(11), 1724–1736. https://doi.org/10.1080/13501763.2019.1668045

Mühlböck, M. and Tosun, J. (2018), 'Responsiveness to different national interests: Voting behaviour on genetically modified organisms in the council of the European Union', *JCMS: Journal of Common Market Studies* 56(2), 385–402.

Mühlböck, M. and Yordanova, N. (2017), 'When legislators choose not to decide: Abstentions in the European Parliament', *European Union Politics* 18(2), 323–336.

OECD (2006), *Applying Strategic Environmental Assessment. Good Practice Guidance for Development Co-operation*, Paris: OECD Publishing. https://www.oecd-ilibrary.org/content/publication/9789264026582-en

Pennock, J. R. (1952), 'Responsiveness, responsibility, and majority rule', *The American Political Science Review* 46(3), 790–807.

Radaelli, C. M. (2017), *Technocracy in the European Union*, London: Routledge.

Rauh, C. (2016), *A Responsive Technocracy?: EU Politicisation and the Consumer Policies of the European Commission*, Colchester: ECPR Press.

Rauh, C. (2019), 'EU politicization and policy initiatives of the European Commission: The case of consumer policy', *Journal of European Public Policy* 26(3), 344–365.

Rauh, C. (2021), 'One agenda-setter or many? The varying success of policy initiatives by individual Directorates-General of the European Commission 1994–2016', *European Union Politics* 22(1), 3–24.

Reh, C., Bressanelli, E. and Koop, C. (2020), 'Responsive withdrawal? The politics of EU agenda-setting', *Journal of European Public Policy* 27(3), 419–438.

Ringe, N. (2005), 'Government-opposition dynamics in the European Union: The Santer Commission resignation crisis', *European Journal of Political Research* 44(5), 671–696.

Scharpf, F. (1999), *Governing in Europe: Effective and democratic*? Oxford: Oxford University Press.

Schmidt, V. A. (2022), 'European emergency politics and the question of legitimacy', *Journal of European Public Policy* 29(6), 979–993.

Schmidt, V. and Wood, M. (2019), 'Conceptualizing throughput legitimacy: Procedural mechanisms of accountability, transparency, inclusiveness and openness in EU governance', *Public Administration* 97(4), 727–740.

Schneider, C. J. (2019a), 'Public commitments as signals of responsiveness in the European Union', *The Journal of Politics* 82(1), 329–344.

Schneider, C. J. (2019b), *The Responsive* Union*: National Elections and European Governance*, Cambridge: Cambridge University Press.

Sloat, A. (2003), 'The preparation of the governance White paper', *Politics* 23, 128–136.

Soroka, S. N. and Wlezien, C.(2010), *Degrees of Democracy: Politics, Public Opinion, and Policy*, Cambridge: Cambridge University Press.

Tallberg, J. and Zürn, M. (2019), 'The legitimacy and legitimation of international organizations: Introduction and framework', *The Review of International Organizations* 14(4), 581–606.

Thatcher, M. and Sweet, A. S. (2002), 'Theory and practice of delegation to non-majoritarian institutions', *West European Politics* 25(1), 1–22.

Thomson, R., Arregui, J., Leuffen, D., Costello, R., Cross, J., Hertz, R. and Jensen, T. (2012), 'A new dataset on decision-making in the European Union before and after the 2004 and 2007 enlargements (DEUII)', *Journal of European Public Policy* 19(4), 604–622.

United Nations Economic and Social Commission for Asia and the Pacific (UNESCAP) (2009), *What Is Good Governance*. Available at: https://www.unescap.org/sites/default/files/good-governance.pdf (Accessed: 14 July 2025)

Van Der Eijk, C. and Franklin, M. N. (2004), 'Potential for contestation on European matters at national elections in Europe', In G. Marks and M. R. Steenbergen (Eds.), *European Integration and Political Conflict* 1 edn, Cambridge: Cambridge University Press, pp. 32–50.

Van Doeveren, V. (2011), 'Rethinking good governance: Identifying common principles', *Public Integrity* 13(4), 301–318.

Williams, D. and Young, T. (1994), 'Governance, the world bank and liberal theory', *Political Studies* 42(1), 84–100.

Wratil, C. (2018), 'Modes of government responsiveness in the European Union: Evidence from Council negotiation positions', *European Union Politics* 19(1), 52–74.

Wratil, C., Wäckerle, J. and Proksch, S.-O. (2023), 'Government rhetoric and the representation of public opinion in international negotiations', *American Political Science Review* 17(3), 1105–1122.

Yordanova, N., Khokhlova, A., Ershova, A., Schmidt, F. D. and Glavaš, G. (2024), 'Curb EU enthusiasm: How politicisation shapes bureaucratic responsiveness', *West European Politics* 48(4), 767–792.

Zhelyazkova, A., Bølstad, J. and Meijers, M. J. (2019), 'Understanding responsiveness in European Union politics: Introducing the debate', *Journal of European Public Policy* 26(11), 1715–1723.

8

NO GREEN CAR

Naïve, balanced, and bleak views of governance

Dan Coffey and Carole Thornley

Global warming, an existential threat to currently existing forms of human society, is simultaneously a crisis for the car because it is in part caused by the car. The car, more generally, exists as a nexus of environmental problems. This chapter begins by reviewing two waves of car criticism, focusing on polluting factors in relation to propulsion mechanisms. It then turns to governance. There is a *naïve* view, in which the problem of the car disappears with technology transition: since internal combustion engine cars burning carbon-based fuels account for exhaust pollutants, the fix lies in saviour machineries, led by a revenant electric car project. Against this, there is a position which holds that the global car fleet, that is to say, the worldwide sum total of cars in use, already massive and continuing to grow, can be maintained neither at its existing size nor on its current trajectory with any hope of more than partially replacing internal combustion engine cars with a different type of car. Furthermore, battery-electric car technology as such is far from free of problems. Contrasting this with the naïve view, we therefore consider in turn two further viewpoints: one we call the *balanced* viewpoint, because it is more alert to relevant factors, and another *bleak*, because it takes fullest account of lust-for-profit and capacities for denial. The naïve view lends itself to thinking about policy packages that can be conceived of in domestic terms, the balanced and bleak extend to the international dimension and resource struggles.

The issues encountered in this chapter are peripheral neither to the problems pollutants pose for the well-being of life on our planet nor to the experiences of marginal or powerless communities hit by an economic juggernaut. The processes leading to the formation of car cultures in the 20th century were integral to the course of social and economic development in advanced capitalist economies. The same applies to the emergence of new industrial

DOI: 10.4324/9781003467694-10

powers rivalling the Western hegemony in the latter decades of that century, running into the world as it exists today. If a naïve take on the car is one which parlays a narrow and evasive prospectus, scarcely more honest than the bleakness proffered by the business-as-usual brigades, the reader must bear the following points in mind when considering the difficulties all proposals to mitigate, reduce, or reverse this unsustainable phenomenon face. A mass of extraction and production complexes, a great network of service industries, patterns of life and standards of personal acquisition, these things and more tie to the car.

Car propulsion: first wave criticism

The still dominant form of modern car might be described thus: a hard shell, enclosing a compartment in which there are seats and control systems, mounted on wheels that are turned by mechanical parts that are themselves moved by forces whose generation hinges on the combustion of fuel to create a hot expanding gas. While this might fairly be criticised for its simplification, as such things go it could be worse. A reader, trying the literature for the first time, will encounter many suchlike efforts.

What this attempts to reference is an internal[1] combustion engine. Fuel is combusted inside fixed-position cylinders, each housing a working part called a piston. For the type most frequently found in cars, a piston moves one way as an air-fuel mix enters the space inside its cylinder (the combustion chamber) via an intake valve; and then back again as the same mix compresses. Ignition creates a hot expanding gas which pushes the piston once more; then an exhaust valve opens, the gas escapes, and the piston again returns. The whole cycle then repeats.[2] In this way each piston, held to a set linear track, moves to-and-fro in a reciprocating motion. This causes another part called a crankshaft, connected by rods, to rotate in a circular motion. It is this latter motion, mediated by a transmission mechanism, which makes the car's wheels turn. A great deal of practically supported research and development has fomented the technology, generating variety in things like size, power, and fuel economy, as well as a plethora of particular design features, to make internal combustion engine architectures dominant. Their principal fuel is gasoline (petrol) distilled from crude oil (petroleum), a naturally occurring substance formed from fossilised organic matter. It contains hydrocarbons: molecules combining hydrogen (H) and carbon (C) atoms.

Environmental consciousness and scarcity panics

A collective ecological awareness with a science-based planetary purview began to coalesce, according to McNeill (2010), in the 1970s. Chlorofluorocarbons, a risk to the ozone layer, became an early case in point. Rapid economic growth in the decades after World War 2, more or less tripling material and energy requirements worldwide over the course of a single generation, also generated waste; and the global population, 2.5 billion in 1950, overtook 4 billion in the 1970s (McNeill, 2010, 266–267). While this last seems small beer today, when UN projections suggest 8.2 billion for 2024, fears that humanity would outrun resources overlapped ecological concern. In this way, planetary-scale environmental consciousness grew up alongside a scarcity panic. McNeill suggests two landmarks for these contrarian impulses. In 1972, the United Nations Environmental Programme was founded. And that same year, the Club of Rome published a best-selling prognosis of ineluctable scarcities: *The Limits to Growth*.

This two-sided angst is readily illustrated for crude oil and gasoline. When crude oil prices soared following action by the OPEC cartel in 1973, fear that supplies might permanently peak – 'peak oil' – began to take root in public consciousness. In the mid-1950s, geologist Marion King Hubbert had predicted that America's crude oil production would peak in 1970, and this turned out to be true (albeit pinnacling at a somewhat higher production rate than Hubbert had thought likely). Now anxiety spread: crude oil production must not only peak everywhere, but do so quickly! Yet at the same time a more expansive environmental sensitivity found disgust in ruinous spillages: oil spills at sea, oil spills on land, oil tankers running aground or sinking. The 1970s in fact ended with a disaster in the Gulf of Mexico, the largest oil spill then recorded, when exploratory underwater drilling operation Ixtoc 1 suffered a blowout (later exceeded by the Deepwater Horizon spillage of 2010, again in the Gulf of Mexico).

The balance of evidence since does not confirm that gasoline, or other oil derived fuel, is liable to run out anytime soon, with improvements in discovery and recovery rates from conventional oil sources on the one side, increased scope for exploitation of unconventional oil sources – heavy oil reserves, in Venezuela; Albertan tar sands, in Canada; coal oils, derived via gasification processes or direct liquefaction – on the other. Shale oil and gas are also a substantial American industry, with outcrops elsewhere. Supply chain threats exist, from wars, embargoes, state policies, and lagging investments, but such resources warn against decisive 'peaking' (for all these points, vis-à-vis peak oil and cars, see: Sperling and Gordon 2009: Chapter 5, cf. pp. 114–130). However, environmental contaminations will continue, a constant threat. Deep-sea mining, relevant to critical minerals and hydrocarbons alike, will add to the toll.

Exhaust pollutants

Awareness that car engine exhaust is not sweet dates to whoever first bent over to sniff. From the vantage point of the 1970s, exhaust pollutants were principally the result of incomplete fuel combustion, high engine heat and pressure, and fuel impurities. Unburned hydrocarbons (HCs), a noxious smog-inducer, and carbon monoxide (CO), bad for respiratory performance, in the first category; nitrogen oxides (NO_x), mainly formed by oxygen combining with nitrogen from the air mixed with the fuel, in the second; and in the third, sulphur oxides (SO_x) and volatile organic compounds (VOCs). Lead in fuels, used as an additive since the 1920s, was a known neurotoxin. In years to follow solutions could be enumerated: catalytic convertors, first appearing in the mid-1970s, to tackle HCs, CO, and NO_x, later supported by systems to improve the quality of engine burn and the catalytic process; improved refining, and emulsions, to tackle fuel impurities. Lead was also phased out, tardily in some countries, as a gasoline additive. However, an intractable problem, particulate matter spewing forth from car tailpipes, as well as road dust and rubber particles,[3] remained as a health-damaging pollutant.

Car propulsion: second wave criticism

Yet the biggest, and planetary-scale, problem took longer to register. Burning gasoline releases energy when its hydrocarbon molecules break-up. Constituent hydrogen and carbon atoms recombine separately with the oxygen atoms in the air-fuel mix. Water (H_2O) forms, together with carbon dioxide (CO_2). If gasoline, in its most conceivably pure form, burns perfectly, this still obtains. It is the unavoidable and intended chemical reaction which provides energy to work the engine. Every U.S. gallon of gasoline combusted, irrespective of the car, releases roughly 20lbs (9kg) of CO_2.[4] Unfortunately, while water vapour in the exhaust is harmless, gaseous CO_2 is not.

This is because of the role CO_2 plays in regulating the earth's energy balance, where it assists atmospheric water vapour in retaining heat from solar energy flooding in. Serious concern about excessively high atmospheric CO_2 dates to the 1950s, when Charles Keeling uncovered through observatory records that rising concentration levels were fluctuating around a much faster than expected upwards trend. A colleague – Roger Revelle – had inferred from empirical measurements that the earth's oceans were less effective at absorbing carbon from the air than previously thought. (For the foundation research on industrial activity and global warming, see: Richter 2010: cf. Chapters 2–5.) Yet decades later, with scientific discourse already entering public discourse, even academic specialists on the car industry – as with oil – were loath to think hard thoughts. An MIT-led survey of the global industry, published in the 1980s, advised thus: 'If other fossil-fuel users find ways to reduce CO_2 emissions, the period for which fossil-fuel use by automobiles is environmentally acceptable may extend far into the future even if CO_2 proves to be a serious long-term problem' (Altshuler et al. 1984, 59). The overwhelming scientific consensus today is that rising CO_2 raises global temperatures. Anthropogenic global warming, far surpassing ultimately manageable threats to the ozone layer, shadows everything: a science-backed planetary scale crisis second to none.

Revenant electric

It is in this context that electric cars, briefly popular in the first decade of the 20th century, have re-appeared, as a saviour revenant-technology. Hybrid electric cars, as opposed to all-electric cars with zero tail-pipe emissions, are not now considered particularly effective in curtailing CO_2 emissions. However, even for an all-electric car to be effective, a 'clean' energy programme is *sine qua non*, since otherwise the electricity to charge and recharge electric car batteries will be generated in ways that simply divert climate emissions from cars to power-station chimneys. The surge in all-electric car use in China, for example, is undermined by its coal power.

As for how all-electric cars propel, to visualise a simple electric motor imagine a stacked series of separately insulated, horizontally oriented, and joined together steel rings, inside each of which are carefully positioned coils of conducting material. This is a fixed position, unmoving, assemblage of parts, called a stator. A rotating magnetic field forms when battery-supplied electric current passes through, causing a rotor slid in lengthwise through the centre of the rings to rotate: a circular motion. This provides the mechanical action which, again feeding through, makes the car wheels turn. However, the great breakthrough has not been at this end of the technology (after all, electric cars existing at the beginning of the 20th century sold well for a time). Rather it lies with advances in chemistry and the development of the lithium-ion battery. The gains thus allowed, in lightness, power, and the number of times a battery can be charged and discharged, are such as to have merited award of the 2019 Nobel Prize for Chemistry.[5]

Battery-chemistry is constantly researched, with a steady stream of laboratory progressions announced almost daily. For example, on work to replace lithium (Li) with alternative elements, such as sodium (Na); on solid state batteries that will eliminate liquid or gaseous electrolytes; or on outré projects like nuclear-powered batteries. Important though some of this must undoubtedly prove, in this chapter we perforce focus on the currently dominant chemistry, for the same reason that we began with gasoline: lithium-ion batteries are, so far, the only real game in town. By the same token, we do not intend to dwell on hydrogen fuel-cell technologies, a perennially delayed hoped-for breakthrough that may, at some point, be finally realised.

Diesels

There have been other responses to the problem of internal combustion engine cars that seek to retain the technology but abandon gasoline. Diesel fuel is a case in point. Derived like gasoline from crude oil, it is denser and harder to ignite. Diesel engines cope by using compression to combust the air-fuel mix rather than an ignition spark. Burning a gallon of diesel actually releases more CO_2 than gasoline. But diesel engines are more fuel efficient, in the sense that less fuel is used for weight-distance travelled, so that substituting in diesel engine cars is a way of lowering emissions overall (reductions, depending on vehicle, varying from 'low' to as much as over 15%). Unhappily, as the evidence on global warming became impossible to ignore, a number of countries opted to adopt measures to promote diesel engine cars. A scandal erupted when it emerged in 2015 that Volkswagen, and later other companies, cheated emissions tests to mask the (NO_x) polluting factor of their diesels.[6] Diesel engines, long favoured in the heavy goods vehicle industry, are also mucky, notorious for problematic particulate matter spewing forth in the exhaust. In a landmark case in 2020, a British coroner ruled that air pollution contributed to a child's death from asthma: nitrogen dioxide (NO_2: an NO_x pollutant) levels in the vicinity of the home massively exceeded international standards, including both EU and World Health Organisation (WHO) (BBC 2020). Exhaust particulates, road dust, and other particles are now linked too, not only to lung and cardio-vascular diseases but also to Alzheimer's[7] disease, instances of which in Britain and some other countries are increasing. While harder to ascribe specifically to a surge in diesel engine cars, there is no doubt left about these pollutants. Battered corporate reputations, and growing air-quality concerns, have since seen moves to discriminate against rather than for diesel engine cars, with phased bans in some cities (including e.g. Athens, Mexico City, and Paris) and partial limits elsewhere.

Biofuels

A quite different response to carbon emissions is biofuel. Ethanol for modified combustion engine fuels is produced from corn in America and sugar cane in Mexico. In Malaysia, by contrast, palm oils form the basis of its industry. Biofuel production is sometimes legislatively mandated, often extending to prescriptive obligations for conventional fuels to be mixed with ethanol or other biofuels. Burning these fuels still produces CO_2 emissions, the aim being rather to achieve some overall net reduction. There are a great many governance issues, including land access and use, biomass diversion and food security, and the implications for biodiversity. However, these ultimately diverse industries remain peripheral to the crude oil core of car fuelling, while any challenge posed to the all-electric car is limited by land availability.

A close to preliminaries

At this juncture, something else needs saying. Cars are, in and of themselves, built from carbon-intensive materials like steel, glass, aluminium, plastics, and synthetic rubbers. Machineries, wherever employed, including transport equipment like trucks used to carry components, are likewise built from materials and run on energy, as too are factories, plants, and storage depots. In fact, every act in procuring fuel, every item of infrastructure, including manufacturing and installing as well as running oil rigs, refineries, pipelines and super-tankers, for crude oil and its distillations, is a carbon emission source. Ditto materials and energy invested in 'cleaner' energy, the electric grid, and electric vehicle charging infrastructures, as well as the production of car batteries. In other words, how car propulsion mechanisms are powered is not the whole carbon story. Much is not tagged against car-use, but is hidden away in the aggregates for emissions attributed to a very wide-ranging set of industries. This means that most accounts of cars, in relation to CO_2 emissions, undersell the problem. Vehicle use, driving extensive road, bridge, tunnel and dedicated parking-space constructions, extends the scope for applications of concrete and other carbon intensive materials.

As to how many cars exist worldwide, this is an informed yet debatable guess. For instance, BNEF (2024, 5) suggests the following. In 2023, there were 1.3 billion passenger cars and 170 million light commercial vehicles, accounting for 53% and 11% of road-transport exhaust emissions of CO_2 respectively: often aggregated for policy purposes, together this accounts for almost two-thirds of emissions. The same source proposes 1.1 billion two-wheeled vehicles, excluding bicycles; 120 million three-wheeled vehicles; and just 3.4 million municipal buses. These account, *inter alia*, for 5%, less than 1%, and 1% of emissions. Medium and heavy road vehicles, estimated at 83 million vehicles, tally up to 30% of emissions (this is the most problematic category for electrification, given vehicle size and power-needs).[8] The figure on cars is likely to be underestimated: 1.3 billion cars are also estimated for 2022. But as a rounded order of magnitude, this provides some rough sense. What is just as important to think about though is that, year after year, the numbers grow.

Governance: a naïve view

With this as a backdrop, consider now a naïve view of governance as this applies to the environmental crisis of the car. Its defining feature is a singular focus on the need for a technology transition that will change the mechanisms which propel cars. But it goes no further, and envisages no other necessities for what must be done. In particular, it sees no case *per se* for addressing the number of cars on roads worldwide, being content with policies that envisage a one-for-one car technology swap. For example, to promote the all-electric car, supported by a cleaner electric grid, consider the following list as illustrative of the categories around which policies tend to cluster:

a Clean energy programmes
b Infrastructure support for public and private charging networks

c Discriminatory tax and duty structures
d Electric car purchase subsidies
e Targeted business grants and loans
f Business advice and technical support
g Public procurement rules
h Domestic production mandates
i Vehicle import rules
j Research support
k Public exhortation

Industrial policy packages, whatever their mix, typically carry multiple aims, and play out differently in different places and at different times.[9] National industrial competitiveness efforts do not disappear with transition policy. Competition for export markets may see different measures adopted for cars sold abroad. Technology support, whether direct or indirect, including via university engineering schools, may skew towards domestically owned businesses, although in countries wholly or largely lacking an indigenous car industry things may be more even handed (an instance involving relatively advanced economies would be France versus Britain, with the first favouring French firms and the second more desperate for inward FDI). This remains true even where there are efforts at policy synchronisation to contain this: for example, European Union (EU) directives on CO_2 emission limits per car per distance.[10] But provided one always bears in mind that industrial policy, like sport, is war by another means, the types of policy that fall into categories like those listed in (a) through (k) above are reasonably indicative of the panoply of measures that domestic policy can pitch: at energy, the electric grid and other infrastructures, at technology, at manufacturers and importers, at business fleets, at the public sector, and, ultimately, at private car buyers.

What this also implies, from the viewpoint of an implicit stance on governance, is a particular preoccupation with policies. Grindle (2010) observes that much of governance literature revolves around *institutions*, and especially public institutions – whether, for example, these can be described as 'fair, judicious, transparent, accountable, participatory, responsive, well-managed and efficient' (Grindle, 2010, 2); *policy* concern, 'what government should do,' then coming alongside (Grindle, 2010, 3). However, it would be hard to argue that institutional analysis greatly informs the naïve view, the primary concern being in fact that states commit to the prescribed policy set. Governments certainly engage with the car and wider automotive industries, albeit when doing so typically placing larger private business interests at the heart of policy formulation. But how these arrangements measure up to broader standards like fairness, transparency, or accountability, can hardly be said to have counted, so far, for very much.

It could be said in response that with car-producing economies spotted across Africa, the Americas, West Asia as well as the Asia Pacific, Western and

Eastern Europe, the question of how different polities compare is very much a secondary issue when focused on the need to substitute-out internal combustion engines. Moreover, with cars found in every country, and indeed most sizeable communities, the argument could perhaps be made that whether or not the institutions mediating policy pass muster on criteria other than their effectiveness, in pressing home the desired goal of substituting one car technology for another, is really neither here nor there. However, this is hardly tenable, and for reasons now considered. In particular, the naïve viewpoint is muted on how physical resources to supply the electric car project are obtained, and thoughtless about limits on their near and medium term availability.

Problems with lithium extraction

We illustrate this by considering lithium, an essential element for lithium-ion batteries. The world's leading extractors at present are China and Australia, together with Argentina and Chile, who share a deposit-rich area with Bolivia, making up Latin America's 'lithium triangle.' Lithium is a highly reactive element found in natural compounds: lithium hydroxide (LiOH) in rock, lithium carbonate (Li_2CO_3) in brine. For rock extraction, ore dug from opencast mines is dissolved in sulphuric acid. Brine extraction, by contrast, removes deposits found in salt flats or salars: holes are drilled, water is pumped and brine evacuated into evaporation pools, to leave residuum containing the sought after lithium compound which is then transferred to treatment plants. China, not the largest producer, has a strong downstream presence.

Lithium extraction pollutes and disrupts. Riofrancos et al. (2023) – a 'Climate and Community Project' comprising overviews, case studies, and an original forecast of the implications of all-electric car conversion for America's lithium needs – describes projections for Thacker Pass, a prospective rock (soft clay) site in North West Nevada. Over a 41-year lifespan, its forecast production yield exceeds 30,000 metric tonnes annually. The anticipated expanse of this opencast mine is extensive: 5,695 acres. Site water-use is expected to equal annual consumption for 15,000 households; while projected waste, whose risk-factors involve water as well as soil contamination, amounts to 354 million cubic yards, over 270 billion litres, of tailings ('tailings' being a general term for waste generated in the process of recovering minerals from ore, with clay a complicating factor for steps then needed to protect at-risk zones) (see Riofrancos et al, 2023, 23–24). Pressure on local water-resources, both from how much lithium mining requires and for what contamination means, is rightly emphasised: 'More than half of global lithium production currently occurs in areas characterised by high water-stress – an issue that will only get more salient as the climate crisis intensifies' (Riofrancos et al, 2023, 21). Environmental damage implies habitat loss for wildlife, including antelopes, grouse, golden eagles and the Nevada state fish, as well as cattle rearing, an aspect protested by ranchers. Unsurprisingly, objections to the mining project listed in this study include pollution, water depletion, biodiversity loss and threats to non-mining incomes (Riofrancos et al, 2023, 21).[11]

However, this is not all. Local Native Americans, like ranchers, are protesting the plan, owing to the site's cultural importance. In fact, almost four-fifths of deposits identified in America sit 'within 35 miles' of reserved lands (Riofrancos et al. 2023: 21).[12] Extrapolating this point, the study highlights potentials for loss of Indigenous sovereignty, cultural threats, and human rights violations arising from expanded lithium mining.

A growing body of writing and comment also exists exploring this issue outside the US. For example, Hernandez and Newell (2022) consider it for salt (brine) extraction in Chile, at the Atacama salt-flats, part of the lithium triangle. The brines take the form of underground waters, not directly disclosed to observers above. Deploying the notion of 'assemblage,' they explore, from the extractor's perspective, the multiple components entailed by and devoted to the project of lithium extraction; with material aspects, institutional aspects, and legitimating discourses. Despite 'mounting evidence that the chemicals and water used for lithium cause large-scale environmental disruption to the salt-flats and surrounding environment' (Hernandez and Newell, 2022, 946), the views of the historical occupiers of the land – the Lickenantay (Licken Antay) people – are marginalised by a state most intent upon maintaining relations with mining corporations, from whom it obtains much of its information on the state-of-play for brine resources, within a national polity characterised by inscribed corruption amongst Chilean elites:

> Behind talk of partnership and green growth we find the appropriation of territory, whereby dominant forms of capitalism preclude possibilities of traditional ways of making use of territory, such as living from live-stock and small-scale agriculture ... mining consumes more water and does not allow other economic activities to be developed ... destruction of the ecosystem that supports peoples' livelihood ... makes them more dependent on it so they are left without options when minerals for mining are exhausted and they lose their jobs.
>
> (Hernandez and Newell, 2022, 959–960)

In this, and in the pursuit of lithium, 'oro blanco,' the white gold, the authors espy repetition, of past 'patterns of dispossession,' and analogy, 'with antecedents in Spanish colonialists' violent search for the sacred gold of the Incas' (Hernandez and Newell, 2022, 959–960).

How much lithium is needed?

As to availability constraints, Riofrancos et al. (2023) use empirical-simulations to explore this question for America by considering projected lithium demand for cars in 2050, if by that point its entire fleet is all-electric (simulation is also made, to which the reader is referred, of cumulative demands for 2010–2050). Assuming a one-for-one substitution of car types, but with *no* diminution in forecast growth in the total number of cars registered for use on American roads, and with zero lithium recycling from old batteries, this study finds that in the year 2050 this future demand, on its own and without regard to demands from other uses[13] or for cars in other countries, stands three times (x3) greater than total global supply at the present date used (Riofrancos et al, 2023, 42). For this statistic, the simulation posits 297.3 million cars in America in 2050 (see Riofrancos et al, 2023, 71–72: Table 9). Forecasts of an unconstrained *global* car fleet by this date exceed 2 billion cars. Using this for argument's sake suggests just how impossible it could be to match lithium demand with supply, because on a like-basis demand for lithium to support all-electric cars would be at least twenty times (x20) greater globally than today's current supply. Even if it were only half this, the pressures on lithium mining would be immense. Although focused on the American case, Riofrancos et al. (2023, 16) do observe a consensus that a supply-chain crunch is immanent sometime in the late 2020s or early 2030s. It is interesting to think about how this might compare with the scarcity panic over oil in the 1970s.

Governance: a balanced view

With these points firmly in mind, turn now to what we are calling a balanced view of the governance issues posed by the car's environmental crisis. By 'balanced' we mean sensitivity to all sustainability factors, or if not all, then at least enough to form the basis of a credible assessment of what these factors might be. Happily, for purposes of framing the Climate and Community Project study – Riofrancos et al. (2023) – already encountered is ideal, because it offers a prospectus which is not only sensitive to problems around resource extraction and limits, but seeks an expansive understanding of how things look from inside the world's biggest car economy, America.

The urban context

To begin with, it is to its authors' credit that they include an urgent case for substantial reorganisation of urban spaces in towns and cities. Cars are tacitly subsidised things, being practically nothing without adequate roads to drive on and places to park, payment for which ultimately falls on taxes only part-borne by car-users, even partially excepting such particular instances as fully self-financing toll roads and car parks, since these would be of limited value without public roads to connect into. (State funding of electric vehicle charging networks actually conforms to this same pattern, support from green campaigners moved by the climate crisis notwithstanding.) Tacit subsidy of privately owned cars knocks at cost-effective public transport: car-enabled urban sprawl reduces population (spatial) densities so that buses, say, make fewer pick-ups per stop over more extended travel-distances (Sperling and Gordon 2009, 6, 8). Urban landscapes shaped by and for cars lock-in dependencies upon car transport for those able to resort to it, leaving distress and socio-economic exclusion to those who cannot.

Low population density in urban space thus creates problems. Riofrancos et al. (2023), again illustrating for the American case, also emphasise intersectionality, highlighting aspects of car-enabled and car-driven city reconstructions often overlooked:

> the United States largely destroyed and rebuilt its cities in the mid-twentieth century in ways that created highly racialized urban geographies characterised by segregation, car dependency and urban sprawl ... injustices and damages wrought by this process have been immense, and the process is implicated not only in the climate crisis but an array of other major social and economic problems.
>
> (Riofrancos et al, 2023, 34)

Pointing to the past role of policy, via subsidies to suburbanisation, highway building, erstwhile renewal projects, and the like, in driving this outcome, they propose a radical restructuring of the same spaces and in a similar time-frame:

> a near inversion of this transformation over a similar time horizon could bring manifold social benefits – as well as reduce harms from minerals mining and likely speed up the timeline of decarbonization.
>
> (Riofrancos et al, 2023, 34)

This call for a rebuilt urban comprised of 'densified' spaces could be supported, in the most ambitious of the scenarios which they consider, by measures like walkways, bicycle lanes, subsidy and share schemes for bikes and e-bikes, car-free streets, zoning restrictions, and congestion charges (Riofrancos et al, 2023, 43), a mix of carrots and sticks. Regard is had to the recovery of 'active' transit modes like walking and cycling.[14]

Battery-pack size and lithium recycling

With impossible demands on lithium resources looming, Riofrancos et al. (2023) supplement the case for urban redesign with two further measures. Observing a sales-weighted average battery size increase of almost threefold since sale of the first 'mass' all-electric car sold in America, the Nissan Leaf launched in 2010, the case is made for reducing 70kWh batteries, posited as a present norm, to 35kWh; thus reducing lithium requirements for the thousands of cells making up each pack (Riofrancos et al, 2023, 35).[15] Recycling lithium from retired batteries similarly has a role to play, although here they note not only patchiness in the data for battery collection and lithium recovery but contrary historical practices where mining for lithium is cheaper than recycling. The impact of increased recycling is not viewed as likely to be large either: 'it will be decades before recycling can meet a fraction of global demand' (Riofrancos et al, 2023, 37).[16] Notwithstanding these supplementary measures, reducing car numbers, in denser urban spaces, is held to be of prime importance, alongside smaller (or) lighter cars and improved recycling.

Governance

Where does governance reside in this scenario? For a country like America, with multiple tiers of formal government, national (federal), state, and municipal, it seems reasonable to urge conforming policies at every level. Riofrancos et al. (2023) more or less do this, while leaning away from the upper tiers as likely to do much. Listing multiple levers available to policy makers – on 'mass transit,' 'land use and zoning restrictions' and 'streetscape planning,' as well as pressure points pertaining to battery size and lithium-recycling – a qualification is introduced on the likelihood of Congress, frozen by 'gridlock' and 'polarization,' acting: their appeal in fact appears most hopefully directed towards decentralised policy and 'community' initiatives (Riofrancos et al, 2023, 44). Yet at the same time, the agendas envisioned do require bringing upper tiers on board: 'mining must be central to US transportation decarbonization scenarios,' extending to 'trade agreements' made to incorporate 'supply chain justice,' 'high standards for labor, human and indigenous rights,' and all necessary environmental protections (Riofrancos et al, 2023, 44).

The width of conception on display easily surpasses a naïve view of governance, one-sidedly focused on policies to transition alternative car technology but largely oblivious not only of global resource constraints but to precisely these issues. And it is difficult to disagree with a call for upheld community rights, environmental protections, water security, and democracy in decision-making, all laudable and present. Yet even merely stating this brings home just how intractable things are likely to be.

Governance: a bleak view

If this is indeed so, what will the near-term hold? A quite likely scenario will be that by 2050 the number of all-electric cars used globally is much larger than today. However, these cars will still constitute the minority technology, the reductio *ad absurdum* of this situation being a world inflicted by much the same pollution factors currently associated with internal combustion engines, but with the added burden of a second mammoth fleet made up of all-electric cars co-existing alongside. The reason why this is plausible will lie in a failure to curtail growth in car numbers worldwide.

Commercial practice in capitalist economies orients towards profit rather than other considerations, absenting contrary, enforced, and binding regulation. On questions of outcomes for car size, consider the following. In the 1970s and early 1980s, OPEC-boosted oil prices drove American car buyers towards fuel economies: smaller cars bought, diesel engines preferred to gasoline. But after oil prices fell back, demonstrating what Bannock (1973, 226) once called automotive elephantiasis, a corporate taste for selling 'big' as 'best' resumed. By 2008, for instance, the Honda Accord, as sold in the States, weighed 75% more than it had in 1976 when it broke through as a successful Japanese import predicated on fuel economy; and in these later decades America's domestically owned car makers ramped up on minivans, gorged on pick-ups, and splashed out on SUVs (Sperling and Gordon 2009, 19–21). Since the profit motive applies just as strongly to all-electric car production, attempts to reduce battery size by pressing for smaller or lighter cars will be fought tooth and nail. And similarly if requirements to use more recycled lithium raises manufacturer costs.

Tough as this fight would be, it is currently a moot point. Today's pro-industrial nationalism, dismissal of electric car mandates, tariff-games, aggressiveness on claims to world resource shares, conflict with China, none of this is consistent with an America ready for a balanced response to the environmental crisis of the car. But even without this, and battery-size and lithium recycling aside, the corporate legal and lobbying objections that would be brought to bear, by Big Car and Big Feeder industries, supported by workers anxious for jobs and outraged car-lovers, on any policies strong enough to seriously impinge car use, car production, and car sales would be immense.[17] Looking around the world, there is little to suggest sufficient differences elsewhere.

In cases like the Lithium Triangle, pressures for rapid lithium extraction are liable to increase rather than decline, making it more difficult for Indigenous and other communities experiencing detriment to negotiate solutions. In Chile, for example, in a study already noted, Hernandez and Newell (2022, 954) describe a 'centralist' state apparatus sitting at the apex of a chain of national, regional and local (rural) tiers, connected to mining corporations which have the resources to manage a presence both around the centres of national decision-making and in the mining areas. For a political economy of this kind, extractivism is liable to prevail over objections. Another study of the region, by Villalba-Eguiluz and Etxano (2017), informed by Ecuador but attentive to governance in the wider region, suggests this is not a unique problem. (Other regions are afflicted too, sometimes appallingly so, as in the Democratic Republic of Congo, where Cobalt (Co) extraction, a metal used in some lithium-ion batteries, has produced multiple horrors, including sexually abused trafficked child-workers.)

But what about a round-about route, tackling car dependencies by pushing policies to reform and reorganise the urban spaces? Urban sprawl cuts down the green spaces around towns and cities, destroying ecologies. Enabled by car use, it increases the extent of road-building, and the share of what American critics call pavement[18] in town and city development: not only roads, but sidewalks, parking spaces, and other areas paved over to service cars, including garages, repair shops, and fuel stops. Once paved over, city architectural heritages are also lost, gone forever. In some particularly unfortunate cases, extensive pavement creates heat deserts. In Phoenix city in Arizona, summertime temperatures are rising: asphalt coated surfaces show temperatures going as far as 180°F – water boils at 212°C – or 80°C.[19] Heat-stroked cities of this ilk now issue warnings. At particular risk: the poor, homeless, children, animals.

Yet when personal transit is sufficiently commoditised for car-culture to become a relevant appellation, daily living is vested in pavement. Clearing out ground space for car-use may disadvantage working class neighbourhoods (its corollary being that poorer children are the children who disproportionately choke on pollutants), and prejudice help determine whose housing leaves them most vulnerable to early death; but amidst this, the paved-over spaces hold jobs and businesses, a great web of car-linked opportunities, small and large, spreading over from private undertakings into the offices of the organs of local democracy, in municipal government. Car-related services and fuel stations, employing more people directly than car makers do worldwide, quite literally live on, as well as off, pavement. Catering and leisure too are car-dependent worlds when the car becomes the fulcrum of mobility. To which one should add the disdain many people have learned for living in mass apartments.

The bleak view of governance

Ultimately, cars writ large have never been governed. An imperfect analogy, instructive because of its imperfections, is this. Mkandawire (2007) observes how the ideal of good governance, pressed by African scholars wanting to insist on the importance of democracy, citizens' rights and social inclusivity in policy formulation, was captured and suborned by Western actors who turned it into a rhetorical device to help project their own preferences and cover up their mistakes – thereby facilitating policies by agencies like the World Bank that run roughshod over democracy. From this perspective, governance is a putty-word, moulded by force majeure. The first imperfection, in looking to adapt and apply this to cars, is that there are no powerful agencies seeking to over-rule national preferences favouring car-use as a transport mode. While in the past a great many countries (including post-independence India, early Communist China, and the Soviet Union) deliberately curtailed resources for passenger cars, this has largely been untrue of the West, outside the exigencies and aftermaths of world wars. As the Western model has spread, the limits that existed have vanished. And the second imperfection is this: where cars are aspirations, democracy colludes.

A closing observation

There is another development in the car world. Self-driving, utilising laser scanning technologies enabling autonomously guided vehicles to frame 3D-maps of their surrounds, embedded in an internet of things: cars talking to cars, cars talking to infrastructures, cars talking to people. The governance issues, for this largely ungovernable technology set, are more than several, and as such lie well beyond the scope of the present chapter to consider. Yet insofar as the themes of this chapter are concerned it points, in all likelihood, to a doubling-down on an existing state of affairs: the sweep of the car park will extend, since cars, using more energy, can search for parking spaces miles away before returning to user, and congestion pressures on car numbers in urban spaces will become a tiny bit less severe. Pundits who make self-driving networked cars synonymous with a 'goodbye' to the combustion engine are wrong, and weirdly so. Whether the TV watcher[20] who decants from couch to self-driving car to watch more TV is living the dream is best left to couch potatoes.

Conclusions

We thus conclude by summarising our intended scheme of exposition. Following preliminaries to convey how car technologies link to propulsion, with a hint of the resource demands that are brought in their wake alongside environmental problems, not least of which is exhaust pollution, three perspectives on governance are considered that differ sharply on what 'governance' might mean. The first is narrow, a world of targets and policy goals, driven by faith in the viability of technology as the means of solving car-related problems, even global warming. The second is expansive, and in the course of offering a more balanced approach, sensitive to the never-ending resource needs of cars, how these are obtained, and what the near-term limits are, captures at the point of policy recommendation all that is missing in the first perspective. The third plunges into the bleak realities of what is unlikely to happen, which is to say more or less any of the intelligent recommendations set forth in the second perspective. At this point, the thought proffered is that governance, applied to the overpowering trajectories of the car as a technology that reorganises the world around it, and to a degree that makes it a mighty part of the warp and weft of 'development,' is illusory. Money can be made, communities can be broken, people can car-aspire, and even get to drive, but a great challenge for the ages will not be managed. And there it ends.

Notes

1 Steam cars, deploying *external* combustion engines, a likely industry prospect in the 19th and first decade or so of the 20th century, are just about covered by the same description.
2 This describes an Otto-cycle: the piston moves, or 'strokes', four times through the cycle.
3 Dust and rubber particles are not mitigated by electric vehicles.
4 To convert to an imperial gallon, favoured in Britain, add another 20%.
5 For particular details concerning the lithium-ion battery and its development, the reader is best directed to the awarding committee's website. The recipients of the prize, John B. Goodenough, M. Stanley Whittingham and Akira Yoshina worked on this battery at different times and in different contexts. In general, batteries possess an anode, which releases electrons, and a cathode, which collects them, separated by an electrolyte. They produce current when circuit-connected.
6 Nitrogen oxides, like particulate matter, are harder to tackle with diesel engines. Another issue to emerge from the scandals, largely ignored, is how CO_2 emissions checking can be cheated.
7 See, for example, Calderón-Garciduenas and Villarreal-Ríos (2017).
8 Heavy vehicles *not* counted here, because not classed as road-transport, include myriad things e.g. combine harvesters (agriculture) and amphibious landing vehicles (military). Not all automotive products move on wheels: tanks, like cranes, some tractors, bulldozers and diggers, use tracks.
9 Anyone interested has no choice but to delve into country specific material. For how the naïve view of governance looks in the British case say, see Coffey and Thornley (2015). The most important development since the period covered has been a prospective ban on sales of new cars equipped with internal combustion engines, with the intended date recently shifting from 2030 to 2035.
10 This is described by the EU as a Corporate Average Fuel Economy (CAFE) standard. Note that standards of this kind, sparked in the 1970s by high oil prices, predate carbon concerns.
11 Illustrating how the climate emission toll for cars implicates many industries, Riofrancos et al. (2023: 22) note that mining accounts for 4–7% of emissions worldwide. A small share of this today will be from lithium extraction for cars, but the amount will rise. More generally, a number of mining industries are essential to the car industry, not least coal (for steel) but in other areas too.
12 The story of oil in relation to how Native American rights are protected is unedifying: one could begin with the huckstering and murdering of the Osage in the 1920s, and continue through to how protests led by Sioux against a pipeline impinging burial sites and water resources were handled in 2016.
13 Lithium has other uses of long-standing, like ceramics, glass-making and drugs. But lithium-ion battery use has only recently become ubiquitous. Bogost (2024) notes smart phones and watches, laptops and tablets, toothbrushes, can openers, pepper grinders, nose hair trimmers, leaf blowers and adult sex toys.
14 For transit between cities Riofrancos et al. (2023) advocate investment in regional railways.

15 The actual sales-weighted average for America is placed at 76.75kWh. This exceeds a 70kWh norm, but the numbers of car sold close to this 'norm' is large and taken as representative.
16 In the unlikely case that battery lithium is being recycled at a 98% rate, this study finds that demand in 2050 in America for cars is still twice (x2) current global annual production.
17 How hard commercial interests fight their corner is intimated by the American Corn Growers Association support for action against the Biden electric car mandate, to defend corn ethanol.
18 In British use 'pavement' refers solely to what Americans call sidewalks.
19 On hot America, CNN (2023) begins with a hospital burns unit.
20 Or on-screen game player ... on-line shopper ... social media addict ...

References

Altshuler, A., Anderson, M., Jones, D. and Womack, J. (1984) *The Future of the Automobile: The Report of MIT's International Automobile Programme*, Cambridge, Massachusetts: The MIT Press

Bannock, G. (1973) *The Juggernauts: The Age of the Big Corporation*, Harmondsworth, England: Penguin Books, Ltd

BBC (2020) 'Air pollution death ruling: What comes next?' *Claire Marshall BBC News*, December 17, 2020 https://www.bbc.co.uk/news/science-environment-55352247

BNEF (2024) 'Electric Vehicle Outlook 2024: Executive Summary', *BloombergNEF*, June 12, 2024 https://assets.bbhub.io/professional/sites/24/847354_BNEF_EVO2024_ExecutiveSummary.pdf

Bogost, I. (2024) 'Lithium-Ion Batteries Have Gone Too Far', *The Atlantic*, July 31, 2024

Calderón-Garciduenas, L. and Villarreal-Ríos, R. (2017) 'Living close to heavy traffic roads, air pollution and dementia', *The Lancet*, Vol. 389 (100070), pp. 675–677

Coffey, D. and Thornley, C. (2015) 'Industrial policy: A Green Agenda', in D. Bailey, K. Cowling and P. Tomlinson (eds) *New Perspectives on Industrial Policy for a Modern Britain*, Oxford: Oxford University Press, pp. 395–420

CNN (2023) 'It's so Hot in Arizona, doctors are treating a spike of patients who were burned falling on the ground', *Jen Christensen CNN*, July 24, 2023 https://edition.cnn.com/2023/07/24/health/arizona-heat-burns-er

Grindle, M. (2010) 'Good Governance: The Inflation of an Idea', Faculty Paper Working Series: Harvard Kennedy School, pp1–21. Also in B. Sanyal, et al. (eds) (2012) *Planning Ideas That Matter: Livability, Territoriality, Governance and Reflective Practice*, Cambridge: MIT Press, pp. 259–282

Hernandez, D.S. and Newell, P. (2022) 'Orco blanco: assembling extractivism in the lithium triangle', *The Journal of Peasant Studies*, Vol. 49 (5), pp. 945–968

Hubbert, M.K. (1956) 'Nuclear Energy and the Fossil Fuels', in *Drilling and Production Practice: American Petroleum Institute*; a Shell Development Company Publication, no. 95

Mkandawire, T. (2007) 'Good Governance: The Itinerary of an Idea', *Development in Practice*, Vol. 17 (4–5), pp. 679–681

McNeill, J. R. (2010) 'The Environment, Environmentalism and International Society in the Long 1970s', in N. Ferguson, C. S. Maier, E. Manela and D. J. Sargent (eds)

The Shock of the Global: The 1970s in Perspective, Cambridge and London: The Belknap Press of Harvard University Press, pp. 263–278

Richter, B. (2010) *Beyond Smoke and Mirrors: Climate Change and Energy in the 21*st *Century*, Cambridge: Cambridge University Press

Riofrancos, T., Alissa K., Dayemo, K.K., Haugen, M., McDonald, K., Hassan, B., Slattery, M. and Lillehei, X. (2023) *Achieving Zero Emissions with More Mobility and Less Mining*, Climate and Community Project (CCP), published January 2023 https://climateandcommunity.org/research/more-mobility-less-mining/

Sperling, D. and Gordon, D. (2009) *Two Billion Cars: Driving Towards Sustainability*, New York: Oxford University Press

Villalba-Eguiluz, C.U. and Etxano, I. (2017) 'Buen Vivir vs Development (II): The Limits of (Neo)Extractivism', *Ecological Economics*, Vol. 138, pp. 1–11

9

CONCLUSION

Maria Gavris and Jessica Savage

Introduction

The primary goal of this book was to present a range of alternative perspectives on what constitutes good governance for sustainable development, enriching an underexplored debate and challenging the dominant narratives in the current good governance agenda. We sought to explore two key questions: *What makes governance good?* and *What are the main obstacles to achieving it?* To do so, we draw upon the diverse case studies examined by our contributors, across various levels of governance, geographical regions, and policy areas, highlighting the contextual nature of 'good governance.' Our central argument is that while good governance is essential for sustainable development, the concept's analytical depth must be rediscovered and applied to critically reshape our current development model.

Before turning to the specific lessons around the elements that may constitute or hinder good governance, it is important to acknowledge and highlight a running thread throughout this book: the highly contested and subjective nature of ideas in the field of sustainability. Both 'good governance' and 'sustainable development' are often presented as neutral, unambiguous, universally accepted terms. In reality, they are none of these things. These concepts are inherently political and value-laden, and when considered together, complexities only deepen. If the aim of sustainable development itself is ambiguous and disputed, then any notion of what constitutes 'good' governance for sustainable development will inevitably vary, potentially leading to radically different interpretations. However, this plurality is often overlooked. The current good governance agenda serves specific interests, yet it is often presented under a veneer of objectivity, as if widely accepted and beneficial for all. Its

DOI: 10.4324/9781003467694-11

technocratic and ostensibly apolitical framing is itself deeply political, as it sidelines the participatory processes we consider central to sustainable development. There is actually a dialectical relationship between governance and sustainable development, in the sense that conceptions of development shape governance priorities, while different ways of doing governance support different types of progress and development outcomes.

As such, as we highlighted in our introduction, the issue with good governance debates is not the absence of theory *per se*, but rather the dominance of implicit theories and ideologies embedded within the prevailing good governance agenda. A singular version of good governance is taken to be the norm, crowding out alternative perspectives. We emphasise the need to question the ideological foundations behind the dominant good governance indicators. The current agenda largely serves the interests of the Global North and global neoliberal institutions. But while it is essential to recognise the pro-growth and neocolonial underpinnings of this agenda, this does not mean that the broader idea of good governance should be dismissed altogether. Its value lies in drawing attention to the political and administrative dimensions of development that were long neglected. Ultimately, we argue that good governance must be reconceptualised as both deeply political and inherently contested, requiring critical engagement with power, history, and alternative visions of development.

What makes governance good?

In the introductory chapter, we argued that the question of what makes governance good does not lend itself to a singular, universal answer, but one that is instead context-dependent, socially and culturally constructed, and multifaceted. While this remains true, several common patterns emerge across the chapters in this volume.

This volume affirms the proposition that good governance must be decolonial. In the introduction, we outlined how the mainstream discourse on good governance – far from being neutral – is deeply embedded in a neocolonial paradigm, perpetuated most notably by the World Bank. In the mainstream agenda, 'development' is synonymous with Western capitalist ideals, and 'governing unfolds through the imposition of development as a universalized narrative and project of civilization' (Icaza, 2022, 45). Dominant notions of good governance reveal the enduring footprint of colonial rule. Yet, as we have also emphasised, the concept of good governance was originally critical and anti-imperialist at its core – and there remains much worth reclaiming in this discourse.

In general terms, decolonial governance involves fostering political autonomy and self-determination for indigenous communities; encouraging solidarity and cooperation among Global South countries as they challenge conventional development models; and fundamentally rethinking how decisions are made, who makes them, and the values that guide those decisions, particularly in relation to our shared resources and collective future. Chapter 5, by Vanessa Burns, delved further into what decolonial governance entails in practice, with application to environmental governance. While this chapter argued that there are moral, prudential, and pragmatic reasons to pursue decolonial governance, it also acknowledged a fundamental limitation, namely that colonialism cannot, strictly speaking, be undone. As such, this chapter proposed a more realistic approach, not decolonisation in the literal sense, but rather a *supracolonial* project, one that seeks to move above or beyond colonialism. It illustrated this through the examples of iTaukei hybrid governance in Fiji, and women's custodianship of spring systems in Timor-Leste, highlighting indigenous knowledge systems and relational approaches to nature that challenge prevailing orthodoxies, offering a model of governance that transcends colonial frameworks. Other chapters in this volume have also explored alternatives to the World Bank's development model and its associated good governance prescriptions. For instance, Chapter 3, by Maria Gavris and Jason Heyes, discussed degrowth as a compelling alternative to Western capitalist development models – one which replaces economic growth as a central policy goal with values such as sufficiency, care, and the preservation of existing resources. Recognising plurality in terms of different visions for the kind of society we seek to build and the pathways for getting there is central to the decolonisation of governance.

Throughout this volume, with contributions from authors across diverse disciplines (Development Studies, Economics, Environmental Science, Geography, Management, Politics, Public Policy) and various policy areas, we have also emphasised that good governance for sustainable development is inherently interdisciplinary. This is because, contrary to the mainstream model of development, focused on economic growth, we see development as situated within a web of interdependence – between biotic and abiotic factors, ecosystems, and societies. This perspective challenges traditional disciplinary hierarchies and calls for a more holistic approach. As a result, governance for sustainable development must be understood as multifaceted, requiring insights from diverse disciplinary backgrounds and theoretical frameworks to grasp its complexity. This ties closely to the earlier discussion on decoloniality: the enduring legacies of colonialism and their contemporary expressions are themselves multidimensional. No single discipline can adequately capture their full impact on development and governance. An interdisciplinary approach is therefore not just beneficial, but essential.

Linked to this, good governance must also be inclusive and intersectional, recognising how identity shapes individuals' and communities' experiences of development. Intersectionality holds that social identities – particularly gender, race, and class – cannot be understood in isolation from one another. Instead, these identities intersect to produce distinct material realities and experiences of exploitation and exclusion. Structural inequalities are embedded in governance systems, perpetuated through laws, policies, and bureaucratic practices shaped by broader systems of oppression. Accordingly, the 'goodness' of governance is a relative concept; governance must be judged in terms of how it affects people across different social groups, identities, and locations. This holds true whether considering gender-sensitive policies, as explored in Chapter 4 by Marta Guerriero; environmental policy that is attuned to the needs of local communities, as examined in Chapter 2 by Jessica Savage; or degrowth policies that are attentive to gendered and racialised labour patterns, as discussed in Chapter 3 by Maria Gavris and Jason Heyes.

Relatedly, good governance must be rooted in communities and democratic processes. Good governance entails not only hearing but truly heeding the voices of those who stand to be the most affected by certain policies and programmes. As argued by Marta Guerriero in Chapter 4, good governance requires treating people as agents of change rather than passive recipients of policy interventions. This requires engaging with local leaders, groups, and institutions, while also critically examining local power dynamics to understand who is excluded and why. Such an approach demands participatory, co-creative governance processes, as outlined by Maria Gavris and Jason Heyes in their discussion of decision-making in the world of work in a post-growth scenario in Chapter 3. Similarly, in Chapter 6, by Molly Wimonmat Srichamroen, the importance of co-creating governance objectives was emphasised, with a focus on fostering knowledge-sharing between the public and private sectors to collaboratively address sustainability challenges. In Chapter 7, Anastasia Ershova further emphasised the link between citizen participation, institutional responsiveness, and the broader legitimacy of political systems: governing institutions derive legitimacy by responding to public demands.

There is also a shared understanding across chapters that good governance requires action at multiple levels. Local governance is best placed to respond to community-specific conditions, yet the state remains the central locus of political power. At the same time, transnational challenges such as climate change, human rights violations in global value chains, North-South social justice, or resource dependencies demand global coordination. These different levels of governance are not isolated but interact dynamically. In Chapter 8, Dan Coffey and Carole Thornley argued that governing the car industry requires more than just technological adjustments; it demands a holistic understanding of the car within its broader global context. This includes examining where and how resources are extracted, and considering who bears the greatest burdens, both as a result of the extraction process and from the wider impacts of car use. In the same vein, in their chapter on labour governance, Maria Gavris and Jason Heyes argued that the role of labour in achieving sustainable development must be redefined on a global scale, but that implementing such a transformation would then rely on the state to redistribute income and allocate and reward different types of work, as well as on local communities to interpret and adapt these large-scale transitions to their specific contexts, ensuring that change is both grounded and inclusive.

Obstacles to good governance

As noted above, good governance requires coordinated action across multiple levels. While governance extends beyond the confines of government, existing good governance indicators remain primarily designed for the national level. As Chapters 6 and 7 by Molly Wimonmat Srichamroen and Anastasia Ershova illustrated, translating national good governance principles to subnational and supranational levels poses significant challenges. Even when organisations or institutions develop their own definitions of good governance, they often struggle to implement them effectively, as Chapter 7 demonstrated in the context of EU institutions, and Chapter 6 illustrated in the context of public and private organisations in Thailand. Thinking beyond the current good governance agenda, multilevel governance introduces a distinct and complex set of obstacles. Bottom-up interventions are essential to contest and transform hegemonic notions of governance, but they are not sufficient on their own. Conversely, top-down approaches often lack democratic legitimacy and risk reinforcing existing power structures. The role of the state is also deeply contested. On one hand, the state can be seen as a capitalist instrument that sustains the status quo; on the other, it may be the only actor capable of dismantling structural barriers to sustainability, if it becomes part of a broader counterhegemonic project. Yet not all states have equal capacity: the scope for manoeuvre differs markedly between the Global North and the Global South. Governance rules continue to be shaped predominantly by the Bretton Woods institutions and other actors from the Global North. Histories of colonialism, imperialism, and ongoing forms of neocolonialism have left many Global South countries without the fiscal or institutional room for manoeuvre to support transformative policies. Over the course of this book, we have encountered many examples of people and communities who find themselves at the margins of governance processes. Women, indigenous groups, people working in the informal economy are all under-represented in governance processes and sometimes excluded altogether from governance mechanisms that may protect their interests (e.g. social protection legislation). This is a clear obstacle to good governance.

Illustrating some of these tensions, Jessica Savage highlighted in Chapter 2 the disconnect between global environmental policymaking and the lived realities of local communities, where international priorities often clash with local needs and practices. Chapter 4 by Marta Guerriero illustrated how women's exclusion from decision-making processes leads to gender-focused policies which nonetheless leave the existing gender inequalities intact. Chapter 5 by Vanessa Burns critiqued the 'lack of capacity' trope in environmental policymaking, as a colonial instrument meant to exclude certain types of knowledge, especially indigenous knowledge, from international governance. Chapter 8 by Dan Coffey and Carole Thornley illustrated how the global car industry is governed in a narrow, technocratic manner, which overlooks broader structural issues, leading to human rights violations and environmental damage in 'sacrifice zones' such as the Lithium Triangle in Latin America. Even within the most advanced supranational governance system in the world – the European Union – there remain significant challenges. As Anastasia Ershova showed in Chapter 7, while there is some evidence that EU institutions have gradually improved their responsiveness, it is still unclear which public demands they respond to exactly.

These examples raise critical questions: Which voices influence policy decisions in the global system, and why? Who decides and enforces notions of good governance? Who is excluded from governance, and why? How did these exclusions come to be? How are planetary resources secured? Who has access to them? Under what circumstances? The mainstream good governance model tends to obscure these questions. As Vanessa Burns emphasised in Chapter 5, international governance structures are still deeply colonial and exclusionary, modelled on Western-centric values and epistemologies. Good governance is obstructed by entrenched centres of power and influence and vested interests. The chapters in this book underscore that good governance must confront the historical and ongoing harms committed in the name of 'progress' – including human rights violations and environmental degradation. Any notion of progress must be interrogated in the context of existing power relations.

Ultimately, the core problem in the governance of sustainable development lies in the very definition of 'progress,' of 'sustainable development' itself, which is often tied to a growth-centric ideology. This attachment to a growth-driven model of development renders many so-called technological 'solutions' to climate and sustainability crises ineffective. Governance proposals may appear to advance sustainability, but they rarely interrogate the deeper assumptions underpinning what development means. In this sense, governance for sustainable development becomes an illusion, unable to confront the systemic drivers of environmental and social degradation. As Dan Coffey and Carole Thornley demonstrated in Chapter 8, the car industry illustrates this contradiction vividly. Car ownership is deeply embedded in societal imaginaries as a symbol of progress and modernity, making it a linchpin of development as currently understood. This has resulted in governance frameworks that are effectively powerless – and unwilling – to challenge the dominance of the car as a defining technology of development.

Relatedly, another major challenge in governing for sustainable development is the disconnect between its interdisciplinary ethos and the reality of siloed policymaking. Linked to a narrow conception of sustainability, the current governance model tends to separate policy domains that are, in reality, deeply interconnected. For example, as the chapters by Jessica Savage (Chapter 2) and by Maria Gavris and Jason Heyes (Chapter 3) show, environmental governance and labour governance are often treated in isolation, with labour policy rarely accounting for the broader environmental impacts of different types of work, while environmental policies such as quotas or fishing bans seldom consider the dependence of local communities and livelihoods on these resources. What is needed is a more integrated and holistic approach to governance – one that embraces complexity, centres justice, and recognises that sustainability is fundamentally about redistributing power and resources, rather than quick technical tweaks.

Whether the obstacles to good governance outlined above can be overcome remains an open question, as they stem from entrenched power relations within the global system, the legacies of colonialism, and the prevailing development paradigm. The answer thus depends largely on whether a strong counterhegemonic movement to challenge the status quo can be formed and sustained. Reviving the debate around good governance is not about offering easy solutions, but rather about emphasising the fundamental issues and questions that are largely absent from contemporary mainstream discourse around good governance.

Issues for further research

We believe this book makes a meaningful contribution to the literature on governance for sustainable development and can serve as a springboard for future research. Below, we highlight several areas that warrant further investigation.

Decolonising governance

Decolonising governance is an enormous but necessary task. An important aspect of this process involves embracing contributions from individuals based in or originally from Global South countries. However, as emphasised by Icaza (2022), diversity should not be understood merely as a demographic or nationality-based characteristic. While important, this perspective alone is insufficient for dismantling the coloniality of knowledge. To truly pursue supracolonial governance, as proposed in Chapter 5, also requires 'nurturing a critical awareness about what is erased by a form of knowledge that claims universality and objectivity despite its partiality' (Icaza, 2022, 51). In the context of good governance debates, this would involve further critical research on what has been erased in the process of cementing the current good governance agenda, and de-silencing marginalised perspectives. As Icaza (2022, 51) prompts, we must ask: 'Where do we begin to tell the story of the phenomena of governance and development? Who has been telling that story and who has been listening?'

In the introduction to this volume, we stressed the African origins of the good governance discourse, which was then co-opted by the World Bank and transformed to serve its own interests. It is essential to further explore this history of the concept and potentially uncover other alternative versions of good governance that challenge the mainstream narrative, which have been marginalised or forgotten. De-erasing alternative perspectives on good governance is an important step in decolonising governance. Beyond that, a significant challenge remains regarding translating these alternative governance principles into concrete policy reforms and institutional changes in the context of existing power structures. Therefore, continued work on decolonial governance is needed at the level of both theory and practice.

The methodology of good governance

The dominant methodologies underpinning mainstream good governance frameworks tend to be rigid, technocratic, and overly focused on quantification. These approaches often reinforce the status quo by prioritising measurable outputs over lived experiences and local knowledge. Far from being neutral, such indicators can actually impede genuine reflection on the meanings of sustainable development and good governance. A key strand of the decolonisation critique concerns how Western research often functions as a conservative force aiming to preserve the status quo. This is especially true in the field of good governance. The criticism here is that the mainstream one-size-fits-all governance model implicitly assumes that the Western way of doing things is the only correct way, thereby silencing alternative worldviews. The way forward involves 'replacing the hegemony of Western unitary ways of thinking with more heterogenous systems of understanding the world' (Denscombe, 2024, 232). To advance the field and develop meaningful alternatives, we must adopt methodological openness. Embracing an open ontology – recognising that there is no single definition of good governance – requires openness to diverse ways of knowing, and at the same time, awareness that all knowledge is inherently partial. This 'does not mean that anything goes,' but rather emphasises 'an awareness of the limits of each of our perspectives and promotes an open approach to knowledge' (Icaza, 2022, 52).

Further work is needed to explore and apply alternative methodologies for researching and promoting good governance. Just as the current good governance agenda excludes certain groups, it also marginalises many knowledge systems, thereby perpetuating epistemic injustice. How, then, can we make good governance truly just? Governance must be grounded in the lived realities of those most affected by its outcomes. This demands participatory, inclusive approaches that position local stakeholders not as mere data points, but as co-creators of knowledge and practice. It would be valuable to see more experiential, highly localised case studies, involving constructivist qualitative methods such as participatory action research, with the caveat that even such approaches may not be sufficient to challenge Western hegemony on their own. This is because, as Denscombe (2024, 233) explains, even when research appears to serve 'the interests of those involved,' there remains the critical question of 'who decides whether or not the research will meet the needs of those involved.'

Good governance in the polycrisis

The current global polycrisis has been described as a defining feature of our age (Tooze, 2021). The convergence of ecological, economic, geopolitical, health, and social crises that taken together are more dangerous than the sum of their parts poses serious challenges to good governance for sustainable development. Policymakers are grappling with the difficult question of how to tackle these crises, given their interconnected and self-reinforcing nature, which casts doubt over the effectiveness of current governance frameworks. From a slightly different angle, it can be argued that the polycrisis exposes the limitations of the prevailing development model and existing governance frameworks. As Yuen Yuen Ang (2024) explains, the polycrisis challenges the assumption that Western capitalist democracies represent the inevitable endpoint of human development, with other regions merely needing to 'catch up' through standardised governance reforms promoted by institutions like the World Bank. Instead, she argues, the polycrisis opens space to rethink governance through plural, justice-oriented, and context-specific approaches:

> The polycrisis is paralyzing only for those who are attached to the old order. For those who are not, it offers what I would call a "polytunity" to usher in new paradigms that invert the way we think about the development process, the sources of solutions, and the role of the state.
>
> (Ang, 2024)

Thus, the polycrisis opens new debates on the meanings of development, good governance, and the link between the two. More research is needed to examine how the polycrisis has impacted discussions around good governance and sustainable development, and the extent to which it has challenged the mainstream on good governance. The polycrisis compels us to reconsider not just how governance is practised, but what it is for, and who it serves. It invites further critical reflection on the values embedded within governance frameworks and the methodologies used to assess governance.

References

Ang, Y. 2024. 'Doing development in the polycrisis', Project Syndicate (online), accessed 22 May 2025, available from https://www.project-syndicate.org/commentary/new-economic-development-paradigm-needed-for-climate-change-inequality-pandemics-by-yuen-yuen-ang-2024-11

Denscombe, M. 2024. 'Decolonial research methodology: An assessment of the challenge to established practice', *International Journal of Social Research Methodology, 28*(2), 231–240.

Icaza, R. 2022. 'Decoloniality, governance and development', in W. Hout and J. Hutchison, *Handbook on Governance and Development,* Cheltenham: Edward Elgar, pp. 45–62.

Tooze, A. 2021. *Shutdown: How Covid Shook the World's Economy*. London: Allen Lane.

INDEX

Note: **Bold** page numbers refer to tables; *italic* page numbers refer to figures and page numbers followed by "n" denote endnotes.

For Product Safety Concerns and Information please contact our EU
representative GPSR@taylorandfrancis.com
Taylor & Francis Verlag GmbH, Kaufingerstraße 24, 80331 München, Germany

www.ingramcontent.com/pod-product-compliance
Ingram Content Group UK Ltd.
Pitfield, Milton Keynes, MK11 3LW, UK
UKHW021307260926
475702UK00048B/809

* 9 7 8 1 0 3 2 7 4 1 0 9 3 *